Advances in Polymer Science

Fortschritte der Hochpolymeren-Forschung

Volume 13

Edited by

H.-J. CANTOW, Freiburg i. Br. · G. DALL'ASTA, Milano · J. D. FERRY,
Madison · H. FUJITA, Osaka · M. GORDON, Colchester · W. KERN, Mainz
G. NATTA, Milano · S. OKAMURA, Kyoto · C. G. OVERBERGER, Ann Arbor
W. PRINS, Syracuse · G. V. SCHULZ, Mainz · W. P. SLICHTER, Murray Hill
A. J. STAVERMAN, Leiden · J. K. STILLE, Iowa City · H. A. STUART, Mainz

With 49 Figures

Springer-Verlag Berlin Heidelberg GmbH 1974

Editors

ISBN 978-3-662-15933-0   ISBN 978-3-540-37849-5 (eBook)
DOI 10.1007/978-3-540-37849-5

Originally published by Springer-Verlag Berlin Heidelberg New York in 1974.
Softcover reprint of the hardcover 1st edition 1974
Library of Congress Catalog Card Number 61-642.

Thermal Analysis of Polymers

W. WRASIDLO

Gulf Environmental Systems, Inc., 10955 John Jay Hopkins Drive,
San Diego, California 92121, USA

Contents

A. Scope of the Review

The enormous development of the subject within the last decade is evidenced by the overwhelming number of research papers published, the creation of a number of specialized professional societies and congresses and the appearance of several new journals. Thermal analysis of polymers is too broad in scope to be covered in a single survey. In this report the subject matter is essentially restricted to the problems of the solid state characterization of polymers by thermal techniques, a topic which occupies a central position in Polymer physics. Still, the topic is far too interdisciplinary in nature to lend itself to a coherent line of presentation. Thus the material selected for discussion had to be separated into three individual parts:

1. the glass transition interval of noncrystalline polymers;
2. the transformation of these materials into ordered structures — crystallization —; and
3. melting of polymer crystals,

all challenging and controversial research areas. A fourth section on experimental methods was included separately to avoid distraction during discussion of the principal subjects and because there have been some significant recent advances in instrumental design.

The discussions on glass theories may seem excessive, but are essential in the understanding of the thermal properties of glasses tabulated in Table 2. Finally, a separate discussion on the melting of polyethylene may seem to the nonspecialist unjustified. However, the largest body of definitive studies on polymer crystallization and melting is presently conducted with this polymer, and the authorities speculate that fundamental knowledge gained from this system in the future may translate to other polymers.

As the evidence present itself, it must be admitted that polyethylene is by far more complex than had been suspected when many studies into its solid state behavior were initiated and it will be most interesting to follow any future progress made.

B. Experimental Methods

1. Adiabatic Calorimeters

Measurements of specific heats are generally performed in adiabatic calorimeters. For investigations into the solid state behavior of polymers the nonisothermal adiabatic calorimeter as first described by Nernst (1) is most useful. Basically such a calorimeter consists of an inner sample compartment, an outer heat insulating jacket, a heater and a thermometer. The specific heat of a material is by definition:

$$C(T) = \frac{1}{m} \left(\frac{dQ}{dT} \right)$$

A sample of known weight (m) is equilibrated inside the calorimeter. Then a small amount of heat (ΔQ) is added via electrical heaters and the temperature rise (ΔT) is measured. In order to prevent heat losses to the surroundings, it is important to adiabatically isolate the system by placing it into an evacuated container. Heat losses due to radiation can be eliminated by adjusting the temperature of the container wall to that of the sample via a second heater. The method has been employed to measure specific heats to a precision of less than 0.1% and numerous improvements in design and experimental procedures have been published (2—27). Modifications by Southard and Brickwedde (2) and West and Ginning (7) have extended the temperature range of calorimeters up to 500° C. Dole and students (4—6) have made important improvements by incorporating automatic temperature control and recording components, and Grewer and Wilski (18) claim to have built a system particularly resistent to damage in handling, easy to use and suitable for industrial laboratories. Reference to other modifications can be found in papers by Hellwege and Knappe (9), Sturtevant (10), Tunnicliff and Badley (11), Bartenev and coworkers (12), Tucker and Reese (15), and by Hager (21). Perhaps the most advanced adiabatic calorimeter and one particularly suited for making precise heat capacity

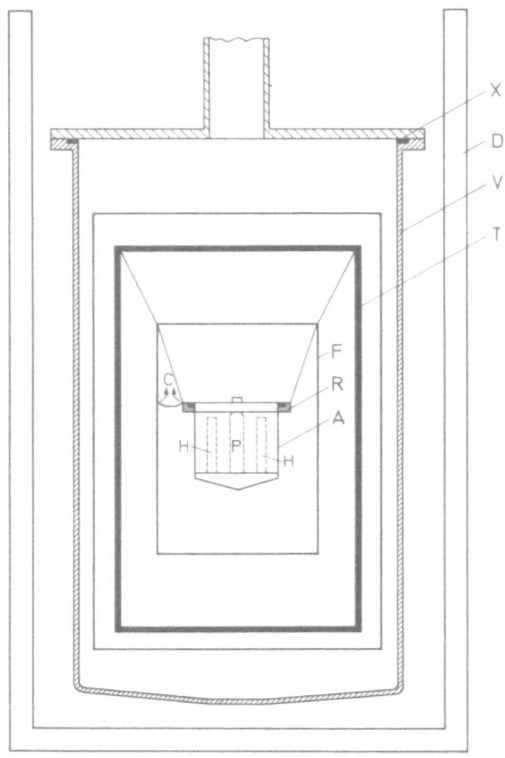

Fig. 1. Automatic adiabatic calorimeter for the study of heat capacities of polymers (14)

measurements of polymers has been described by Karaz and O'Reilly (14) (Fig. 1).

There are several noteworthy features of this apparatus. First, the sample container, the most critical single component of the calorimeter was designed with special consideration given to the low thermal conductivity and low packing density of polymers. The net volume of the sample compartment is only 70 ml and the construction materials used were optimum in terms of thermal conductivity and minimal heat capacity per unit volume of metal. Another unique feature of this calorimeter was the incorporation of a second heat shield (P) placed in between the adiabatic shield and the radiation shield, whose temperature can be automatically adjusted for different adiabatic temperatures in such a way as to keep the power input to the sample container (A) constant. This detail is essential for the automatic control of the calorimeter. With standard materials such as alumina a precision of ± 0.2% of the heat capacity may be attained over a temperature interval from 15—600°.

While adiabatic precision calorimetry will probably remain as the primary experimental technique for the assessment of important thermodynamic properties of polymers in the condensed phase (i.e., Cp, ΔH, ΔS and ΔF) and for accurate determinations of the technically important heat balances in polymeric systems, the method is plagued with problems, severely limiting its wide use as a polymer characterization technique. First, precision adiabatic calorimeters are to the writer's knowledge commercially still not available. There are also problems directly related to the nature of polymeric materials, as for example low thermal conductivity and bulk density, poor packing causing material collapse (i.e., volume shrinkage), clean up problems of calorimeters due to tenacious adherence of samples to the calorimeter walls, degradation of polymer samples during prolonged measurements at elevated temperatures, or temperature drifts due to relatively slow thermal transitions in polymers. Then there exists the problem of calculating the specific heat at constant volume from the experimentally accessible specific heat at constant pressure, a procedure necessary for the interpretation of the vibrational spectra of polymers and for the determination of that fraction of heat which is consumed in volume expansion. In the classical expression

$$Cp = C_v + TV\alpha^2/\chi$$

the compressibility (χ) is required for the calculation of C_v. Unfortunately, so far the compressibility of only a few polymers has been reported (23), thus forcing investigators to make use of empirical relationships (22). But perhaps the most serious drawback to the thermal analysis of polymers by means of classical calorimetry is the slowness of the method which may result in irreversible changes of the sample during measurement. Such changes as a result of the intrinsically high metastability in polymers can influence the results profoundly at higher temperatures, in thermodynamic transition zones. Only very recently (mid '60's) has the extent of metastability in polymers in the solid state fully been realized and at least in part has led to the development of the so-called "fast" calorimeters (45), which will be discussed in a later section.

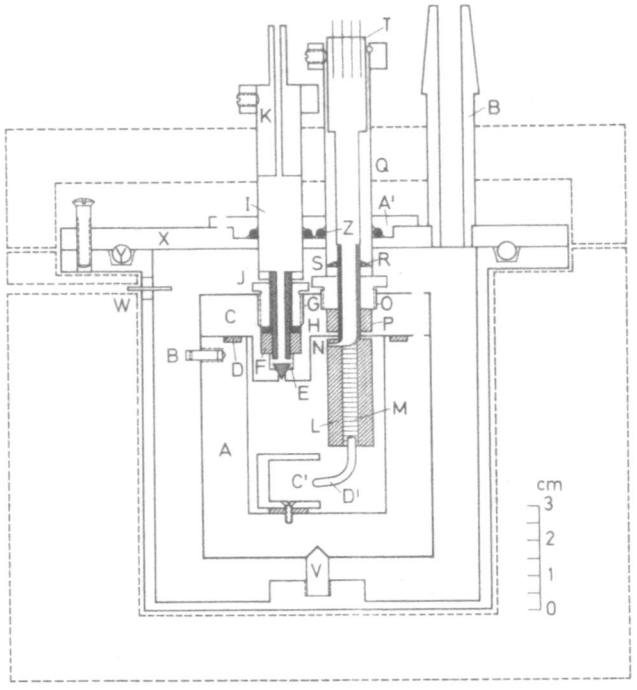

Fig. 2. Sectional diagram of a reaction calorimeter [From Ref. (35)]

Isothermal calorimeters have also been used in polymer studies and primarily in investigations of heats and rates of polymerization reactions (28—37). Shielding and controls in reaction calorimeters are not as critical as in nonisothermal calorimeters used in specific heat measurement since most polymerization reactions are accompanied by relatively large enthalpy changes. On the other hand special attention must be paid in the selection of the calorimeter material for a particular combination of monomers, solvents, catalysts, etc., to ensure inertness of the reactor wall to reaction components. Frisch and Mackle describe an aneroid high precision, semi-micro reaction calorimeter (Fig. 2) usable for a wide range of reaction systems including those in which one component is a gas.

The body of the calorimeter, machined from aluminum (i.e., inert Al_2O_3 layer) has an internal volume of about 50 ml. A thermometer element is connected to the reactor as shown in B. The lid (C) accomo-

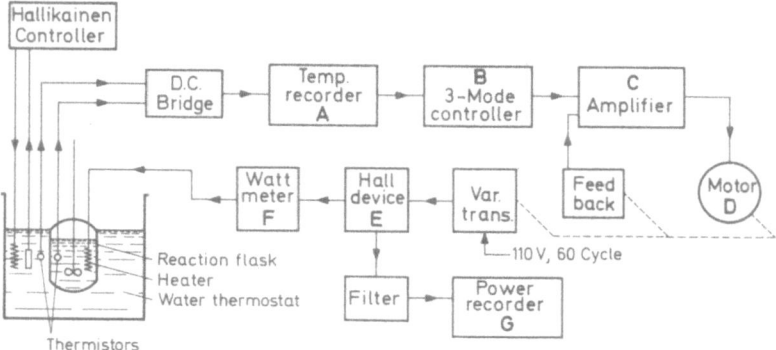

Fig. 3. Schematic diagram of isothermal kinetic calorimeter: *A* Honeywell 10 mV, 10 in. recorder with retransmitting slide-wire, input modified to match dc bridge; *B* Honeywell Electr-O-Volt; *C* Honeywell recorder amplifier; *D* Honeywell chart drive motor; *E* Halltiplier (Scientific Columbus, Columbus, Ohio); *F* Simpson Model 880, 100/200 W; *G* Sargent Model SR. [Ref. (*36*)]

dates a gas inlet needle valve assembly (*E* to *K*), and an ampoule breaker and electrical calibration heater designed as one unit (*D'* to *T*). Surrounding the reactor is the jacket U with a threaded support *V* screwed into the bottom of the wells, thus permitting correct centering of the calorimeter. The authors report a jacked temperature control of $\pm 0.005°$ C over periods of 1—2 h. Heat outputs of the order of 10 cal may be measured with a precision of $\pm 0.2\%$ and temperature-time curves are automatically recorded on a printout counter. The apparatus operates as follows: A few milliliters of monomer under study are sealed in a cylindrical glass ampoule and placed into the perforated aluminum holder *C'*. The second reactant or catalyst is added to the calorimeter in case of a liquid or dissolved solid or in case of a gaseous substance through the needle valve assembly (*E—K*). The ampoule containing the monomer is broken by rotating *Q* so that the steel rod *D'* penetrates it. The reaction starts, and the temperature change is recorded.

Recently Anderson (*36*) described an isothermal kinetic calorimeter capable of automatic instantaneous recording of polymerization rates. A schematic diagram of this apparatus is shown in Fig. 3.

The calorimeter has been utilized in emulsion polymerizations to obtain polymerization rates from temperature time curves, rather than conventional conversion-time curves. By this technique polymerization variables such as electrolyte level, its effects on the critical micelle concentration and kinetic features not otherwise detectable can quanti-

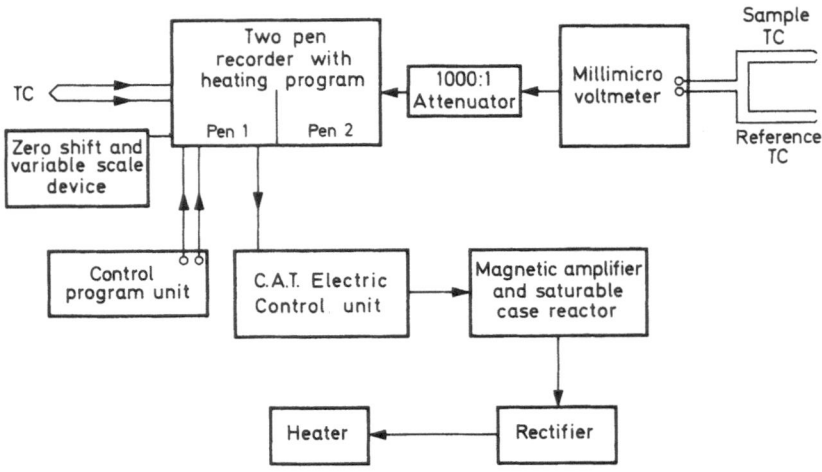

Fig. 4. DTA heating and temperature indicating apparatus [Ref. (42)]

tatively be studied. Other designs of calorimeters of the Calvert-type have been described by Barkalov and coworkers (32) for radiation polymerizations, by Gevorgyan (33) for the polymerization of nylons and by Dworkin in the study of solid state polymerizations (37).

2. Differential Thermal Analysis and Differential Calorimetry

Previous reviews on DTA of polymers have been published in 1964 by Ke (38) and recently by Wunderlich (39). Details on instrumentation and techniques have been reviewed by Smothers and Chiang (40) and by Mackenzie (41). A block diagram of components used most frequently in modern DTA units is depicted in Fig. 4.

The three basic components of a DTA apparatus are a sample cell with heater and enclosure, a temperature indicating device and a temperature control-programming unit. The determination of phase transitions such as glass transitions in polymers necessitates extremely high differential temperature sensitivities. Therefore, recorder preamplification producing a ΔT sensitivity in the range of one μV full scale at a noise level not exceeding 2% is desirable. This sensitivity corresponds to 0.2° full scale of chart when using copper-constantan thermocouples, which is more than sufficient for detecting the presence of secondary thermal processes in polymers.

Another important consideration in the design of DTA equipment is the range of heating rates covered. Important thermal processes in polymers such as reorganization and superheating in crystalline materials, recrystallization of polymer melts, cold crystallization, or volume relaxations during amorphous transitions are time dependent phenomena, often with widely different time constants. Their detection and differentiation from each other and from conventional equilibrium processes can be accomplished by employing a wide range of linear heating and cooling rates. Rates as fast as 1000° C/sec are by no means unrealistic. Unfortunately, most commercial DTA equipment is limited to a range from 0 to about 30° C/min. The temperature range in which relaxations in polymers take place varies from near absolute zero to above 500° C depending upon structure and the molecular nature of a particular process. Therefore it is important for the study of polymers that DTA instruments are capable of operating at linear heating rates over this entire temperature range. Again most commercial instruments do not as yet fulfill this requirement. While conventional DTA has been used extensively (38—44) in determining the temperatures at which thermal transitions occur in polymers, this technique is not suitable for determining the magnitude of associated energy changes to a precision required for thermochemical calculations. Such measurements are conventionally performed in adiabatic precision calorimeters which as

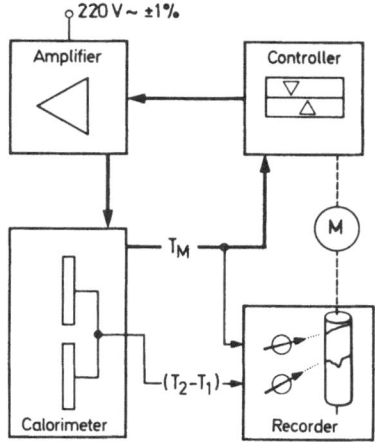

Fig. 5. Schematic diagram of differential scanning calorimeter (DSC) [Ref. (46)]

has been pointed out have their own limitations and drawbacks (22, 45). An attempt to combine the advantages of dynamic DTA (i.e., small sample size and speed of experiment) with those of calorimeters (i.e., precision) was first made by Müller and Martin (46). Their design shown in Fig. 5 is based upon the principle of a differential calorimeter earlier described by Hoffmann (47) in which the heat flow to the sample is registered by means of a temperature difference between the sample and the shield (T_M). However, an additional measurement is made between sample and a reference material inside the calorimeter ($T_2 - T_1$), and the temperature of the calorimeter shield (T_M) is programmed continuously and recorded (i.e., scanned) at linear heating rates (DTA). Heat capacity measurements can be made with a precision of $\pm 2\%$ between -180 and $350°$ C using only 1g samples at linear heating or cooling rates between 0.05 and $1°/\text{min}$.

Another differential scanning calorimeter (DSC) has been described by Watson, O'Neill, Justin and Brenner (48—50). This instrument differs from that of Müller and Martin in that the sample and reference compartment are kept at identical temperatures by means of two heaters, one under the sample holder and the other one under the reference holder (Fig. 6). Reference and sample temperatures are measured by platinum

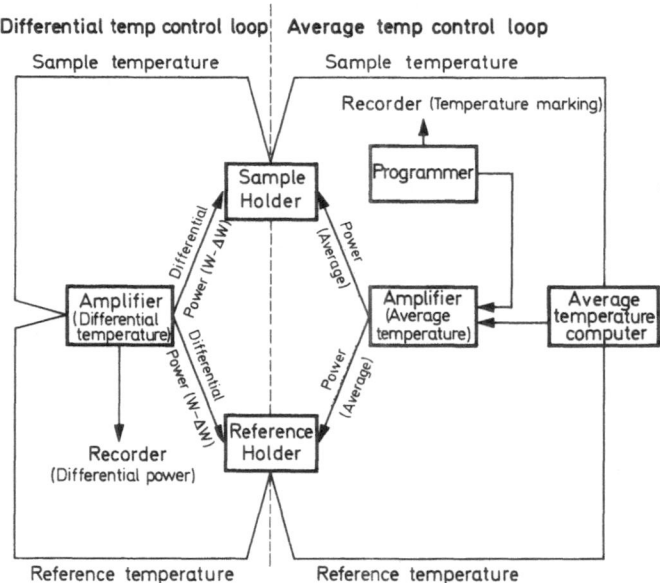

Fig. 6. Block diagram, Perkin-Elmer differential scanning calorimeter [Ref. (48)]

thermometers fed into the differential amplifier whose output then adjusts the differential increment necessary to keep the reference and sample heater power requirements identical. A signal proportional to the differential power output is then recorded as a function of an average calorimeter temperature. Sample weights are between a few milligrams to about 100 mg; heating rates can be varied from 0.5—80° C/min; a temperature range from — 100 to 500° C is accessible.

A series of papers treating theoretical aspects of DSC have appeared (46, 49—55) in which the quantitative capabilities of this technique have been demonstrated. Numerous improvements in design and precision of both the differential temperature and differential power principle have been made (48—65) and two commercial units based on the former[1] and one based on the latter principle[2] are on the market.

Heat capacity measurements on polymers using the Perkin and Elmer Model 1-B DSC[3] were first reported by Wunderlich (66) and by the author (67) employing a DuPont calorimeter cell. Recently the author has also checked the accuracy and precision of the Mettler Model 2000 microcalorimeter using sapphire as a standard and found that its heat capacity could be measured to an absolute accuracy of 2% and a precision of ± 0.5%. So far no heat capacity measurements on polymers have been reported on this instrument. Further improvements in the design and use of these fast calorimeters are surely to come in the near future and should bring these instruments into the range of precision calorimeters.

3. Dilatometry

Thermal expansion measurements provide important information on polymers. For example, they are useful for engineering design purposes, for testing crystallization and nucleation theories, in the study of molecular interactions (i.e., Grüneisen law), for the calculation of thermodynamic (i.e., energy-volume) parameters, for investigations into polymerization kinetics, and for the determination of transition temperatures in polymers. Several reviews on dilatometry have appeared (68—70) and in addition numerous recent papers describing the design, operation and application of this technique to polymers have also been published.

[1] DuPont Calorimeter cell used with either DuPont 900 or 980 DTA, E.I. DuPont Co., Wilmington, Delaware.

[2] Mettler DTA 2000 System, Mettler Instrument Corp., Princeton, New Jersey.

[3] DSC Model 1-B, Perkin and Elmer Corp., Norwalk, Conn., USA.

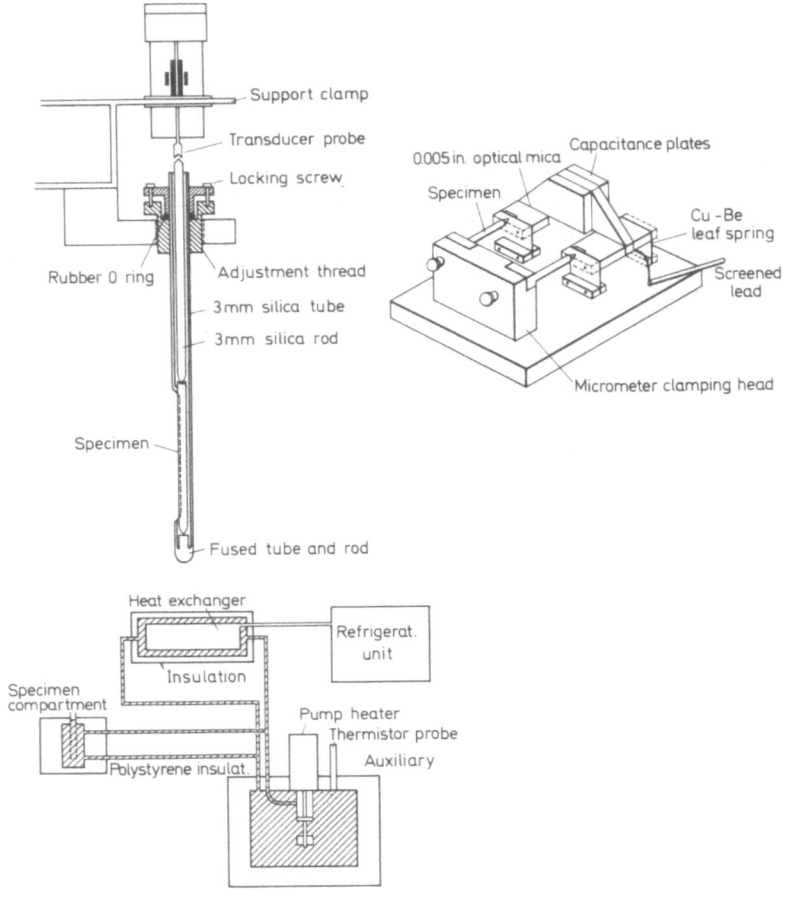

Fig. 7. a Assembly of transducer and specimen holder. b Temperature control system. The specimen tube in (a) fits into the specimen compartment on the left [Ref. (74)]

Schwenker (71) developed a design for use with polymers. A vacuum dilatometer operating up to 1600° C employing optical techniques has been used to measure coefficients of thermal expansion to a precision of 2.2% (72). Evans and Winstanley (73) have described the construction and use of a simple, inexpensive dilatometer for expansion measurements of solids in the temperature range from − 196 to 75°. An extremely sensitive dilatometer has recently been developed by Harrison and

Wilkes (74) for the dilatometric study of vacancies in metals. A diagram of the dilatometer assembly is shown in Fig. 7.

The system employs either a linear transducer probe or a capacitance micrometer probe. In the latter, two specimen can be used, one of annealed and the other quenched material. Differential expansion can be measured by means of a capacitance bridge calibrated using linear expansion coefficients of known metals. The sensitivity of the bridge is sufficient to measure down to 0.001 pF, a sensitivity in expansion of the order of 1 Å can be achieved. Although ultrasensitive dilatometers so far have not been used with polymers, such instruments may in the future prove powerful means in the assessment of solid state volumetric relaxations, which until now have only been observed by indirect methods (i.e., mechanical and dielectric relaxation techniques).

A microdilatometer for the measurement of polymer crystallization rates has been reported by Tung (75). For high pressure operations a dilatometer utilizing a stainless steel membrane as a sensing device has been described by Tsikis and Polyakov (76). Design and operation details of numerous dilatometers for the study of polymerization rates have been discussed by Rubens and Skochdopole (69). Other modern dilatometers useful for polymer studies are described in papers by Hara and Schonhorn (77) for isothermal crystallization work, by Miller (78, 79), by Niezette and Desreux (80) (suitable for radiation polymerization, Fig. 8) by McKinney and Penn (for density measurements of polymers, by Vocks and Crane, (82) and by Duvdevanis (83).

4. Thermooptical Analysis

Visual examination of material during heating or cooling experiments enables one to observe thermal processes such as glass, crystal or melt transitions, the appearance of liquid crystals, colorchanges, etc., which aids in the interpretation of data from other thermal measurements. The oldest and at the present time most frequently used thermooptical analyser is the melting point apparatus. The device has been modernized through the use of temperature programmed hot stages, polarizing microscopes, photocells and chart recorders and a variety of godetry has recently been made commercially available. Miller and Sommer (84) have designed a system for simultaneous DTA measurements up to 1800° C. An instrument capable of DTA, differential thermogravimetry and hot stage microscopy over a wide temperature range up to 1500° C has been described by Wilburn, Metcalfe and

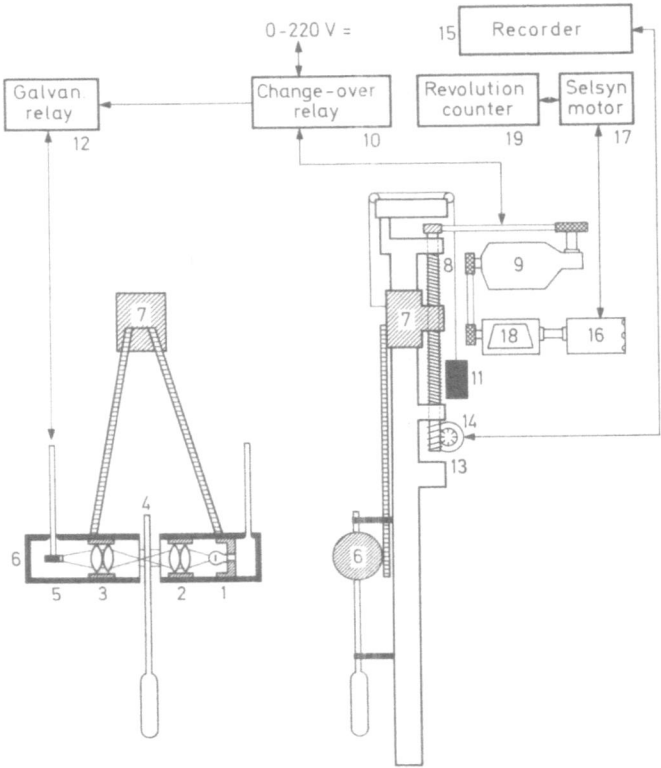

Fig. 8. Schematic diagram of dilatometer: *1* linear filament galvanometer bulb (4 V, 0.5 A); *2,3* biconcave lenses (focal length 21 mm, aperture 20.5 mm); *4* capillary of the dilatometer; *5* Ge photodiode (Philips OAP 12); *6* light source cylinder (length, 70 mm; diameter, 37 mm) and photodiode cylinder (length 125 mm; diameter, 37 mm); *7* mobile block; *8* precision threaded rod (thread 0.5 mm); *9* drive motor (G. K. Keller NHS 12 R, 0—220 V); *10* change-over relay (MBLE R1A/120/KKK); *11* counterbalance weight, *12* galvanometric relay (AOIP RP 221); *13* worm screw; *14* helicoidal potentiometer; *15* recorder; *16,17* selsyn motor; *18,19* revolution counter [Ref. (*80*)]

Warburton (*85*). A novel microscope hotstage, which allows accurate control of heating and cooling rates as well as accurate control in the attainment of temperature has recently been designed by Dixon.

Figure 9 shows a diagram of this instrument. Close control of heating or cooling rates is provided through the use of a thermoelectric unit attached to an asbestos board (insulator). An anodized aluminum

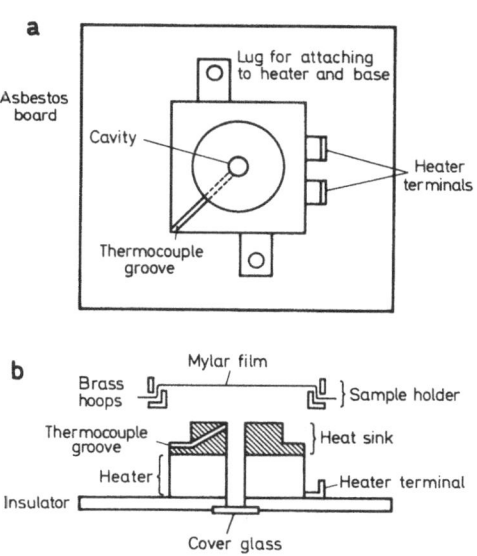

Fig. 9. The microscope hot stage: a plan; b cross section [Ref. (86)]

block was utilized as a heat sink to allow for simultaneous electrical measurements. The sample holder can be moved in such a way as to heat any specific part of the sample. Mettler (87) has made available a commercial instrument for automatic, digital recording of melting and boiling points. The apparatus is based on changes in light transmission which take place during phase changes in materials and operates up to 300° C. The instrument has a disadvantage in that the sample can not be observed visually. Other microscopic hot stages which allow automatic photoelectric monitoring and visual observation of the sample have also been reported (88, 89), including one system for simultaneous hot stage microscopy and X-ray diffraction (90). In a paper by Miller (79), a different thermooptical device, called a thermal depolarization analyzer (TDA) is described. Briefly the equipment consists of a cylindrical furnace containing the sample with windows on both ends, a polarizer and analyzer positioned at each end of the furnace respectively, and a quartz-iodine lamp for illumination of the sample. The output from the phototube (Tung Sol DT 933 A) is fed into an electrometer which in turn is connected to a DuPont 900 Thermal Analyzer. Recordings of transmission of either white or depolarized light as a function of temperature at a given heating or cooling rate can be made.

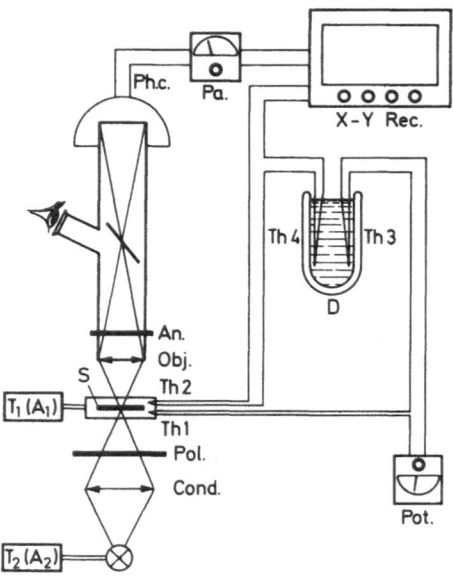

Fig. 10a. Schematic diagram of thermo-optical analyzer (TOA) [Ref. (97)]

This kind of a technique might be useful in the study of mesomeric phases which are difficult to detect by other thermal methods.

Polymers are unique in that they often exhibit anisotropic properties. When light interacts with an anisotropic material, birefringence, the property of having two indices of refraction, is observed. The origin of optical anisotropy may be molecular (i.e., alignment or displacement of chain molecules in a preferred direction) or physical (i.e., boundry regions at crystal-amorphous interfaces, crystal defects, etc.) in nature, and one can expect its magnitude to be reduced by thermal or brownian motions. Therefore, a decrease of orientational or distortional birefringence is related to chain mobility of a polymer which in turn depends on both temperature and free volume. Numerous papers relating to this subject have been published, notably a review of flow birefringence by Tsvetkov (91) and papers by Stein and coworkers (92, 94), Jackson and Longman (93), Grosius, et al. (95), Saotome and Komoto (96), Kovacs and Hobbs (97) and by Shultz and Geudron (98).

In a paper by Kovacs and Hobbs (97), an automatic recording device employing a hot stage, polarizing microscope, photocell and chart recorder is described. A schematic diagram of this thermooptical analyzer (TOA) is shown in Fig. 10a and a miniature stretching jig for producing birefringence in this film sample is shown in Fig. 10b. Runs are made as follows: A sample shield between coverslips or in the jig, is placed in a Latz hot stage mounted on a polarizing microscope (magn. 50—100x). Light is supplied by a 12 volt lamp operated from a transformer T_2 of current A_2. The hot stage is controlled by a transformer T_2 operating at current A_1. Thermocouples Th_1 and Th_2 are used to monitor hot stage temperature. The photocell (Ph) output is fed into the Y input of a Moseley x–y recorder. Birefringence in isotropic samples is produced either by scratching the polymer film with a stylus or stretching the sample in the jig. An updated version of this apparatus employing a Mettler hotstage Model FP 21 with temperature program and control (Mettler FP_2) was described by Shultz and Geudron (98).

5. Techniques of Sample Preparation and Conditioning

It is well recognized that the thermal behavior of a polymer sample depends to a large extent on its thermal history. For example, poorly crystallized polyethylene could contain tie molecules which melt at $-10°$, while folded chain morphologies melt between 80 and 135°, depending upon fold length and defects. Moreover high molecular weight extended chain crystals of polyethylene have an equilibrium melting point at 141° and depending upon heating rates, the interior of such crystals may superheat to 160° or higher (99). While the influence of thermal histories on the molecular, physical and mechanical behavior of polymeric materials has been recognized (and at present is one of the most active areas of research in polymer physics) it is, in the author's opinion, still not fully appreciated. Traditionally, the technique for obtaining an amorphous or metastable crystalline polymer is to cool a bulk sample under some arbitrary conditions and subsequently heat treat (temper) to change the relative amounts of amorphous and crystalline phases to achieve the desired physical properties. Realizing that the production of various phases must be controlled by such factors as the method of heat removal (i.e., conduction to a solid surface, quenching into a fluid, etc.), the cooling rate, the mass of material, etc., the question may be asked whether or not techniques of quenching and annealing could be brought under closer control and the range of conditions be extended. In a recent study (100) it was shown, for example, that highly amorphous polyphenyleneoxide can only be

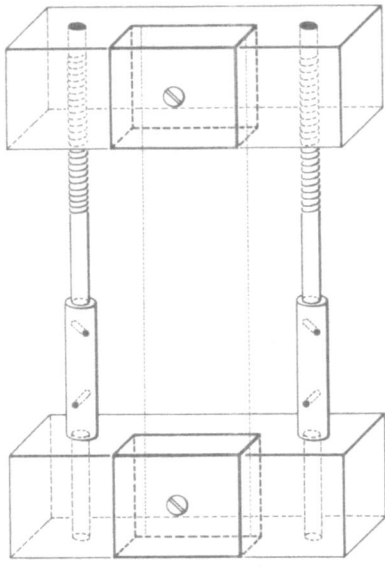

Fig. 10b. Miniature stretching jig [Ref. (97)]

obtained when thin samples (a few microns thick) were cooled at rates in excess of 1000° C/min (see Table 1). Rates of cooling of the order of 20000° C/sec have been attained (101) and it is conceivable that with special techniques this range can be extended to 100000° C/sec and higher.

Table 1. Effect of cooling rate on X-ray crystallinity of poly(1,4-phenylene ether)

Temperature, °C[a]	Cooling rate, °C/min	Sample thickness,	X-ray crystallinity
230	>1000	3 mm	0.27
230	>1000	ca. 10 μ	0
230	100	ca. 10 μ	0.42
230	1	ca. 10 μ	0.45
230	0.1	ca. 10 μ	0.70
112	0.1	ca. 10 μ	0.58
25	0.2% nitrobenzene soln quenched with alcohol	Flocculent powder	0.15

[a] Polymers were held at indicated temperatures for 1 h before cooling. [Ref. (100)]

In another study (*102*) it was demonstrated that under specific conditions single polymer crystals exhibiting equilibrium melting behavior may be grown from the melt. Previously it was believed that single crystals can only be obtained by means of solution crystallization techniques.

In the discussion following a brief account will be given on techniques of rapid quenching and annealing of polymers.

a) Quenching from the Melt. Due to the poor heat conductivity of polymers, special attention must be paid to sample size and geometry in quenching experiments. Small globules of the melt or thin films of polymer melt on highly conductive metal substrate foil (i.e., copper foil) should be used for quenching, regardless what heat transfer mechanism (i.e., conduction, convection, radiation) is considered. The initial sample temperature for quenching depends ultimately on the chemical-thermal stability of the sample. The final choice for selecting the initial temperature has to be made based on previous knowledge of the thermal behavior of a particular material (i.e., glass transition, crystallization and melting temperatures) and will also depend on what morphologies and phases are to be obtained after quenching. For crystalline polymers, their equilibrium melting point is the critical temperature from which the quenching process should be initiated. For amorphous, not crystallizable materials the lowest initial temperature would be the glass transition point and a temperature range in which high fluidity of the polymer occurs would be preferred.

In metallurgy the classical approach to the problem of attaining metastable phases has been rapid removal of heat by convection. Generally a molten sample is dropped into a cooling fluid such as liquid nitrogen. It has been shown that using a cooling gas of high thermal conductivity such as helium at high flow velocities provides a better heat transfer medium than liquids (including liquid helium). In fact, with specimens a few microns thick, cooling rates on the order of 10000—20000° C/sec have been attained (*101*). While the extremely high cooling rates employed in metallurgy are not necessary in preparing completely amorphous samples of many polymers, the technique is very simple and the possibilities of retaining chain conformations typical of polymer melts are so intriguing that it is difficult to understand why so little attention has been paid to rapid quenching of polymers in the past. An apparatus used by the author for quenching of polymer melts by convection is shown in Fig. 11. It consists of a miniaturized capillary extruder which is insulated by means of an asbestos board from a liquid nitrogen bath. The extruder is placed into a hydraulic press, preheated to the desired temperature, a sample (1—3 g) of polymer is placed

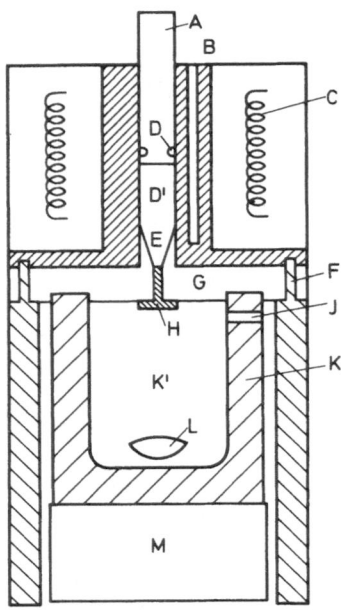

Fig. 11. Apparatus for rapid quenching of polymer melts by convection (*A* piston, *B* thermocouple well, *C* heater, *D* Teflon oring seal, *D'* sample well, *E* disposable copper capillary, *F* locating pins, *G* asbestos board, *H* Teflon disk, *J* gas and overflow outlet, *K* stainless steel Dewar, *K'* liquid nitrogen, *L* magnetic stirring bar, *M* magnetic stirrer)

into the sample compartment, melted, and rapidly extruded into fast stirred liquid nitrogen. In order to prevent the end of the copper capillary to cool off, a teflon disk is placed in front of the orifice (held in place by a coating of oil) and the Dewar flask is connected to the asbestos board just prior to extrusion. Globular samples with diameters of less than 10 μ have been quenched in the apparatus at estimated cooling rates of 10000° C/sec. Samples can be stored in the Dewar flask until used for thermal analysis.

Another apparatus in which heat transfer is achieved by conduction to a solid substrate and one particularly appealing to polymer studies is the "piston and anvil" quenching device. The design, shown in Fig. 12 and based on the principle of an accelerating substrate against a fixed globule was first conceived by Pietrokowsky (*103*).

A small quantity of material is contained in *D* with an orifice at the bottom. The tube is heated by induction through *E* and material is

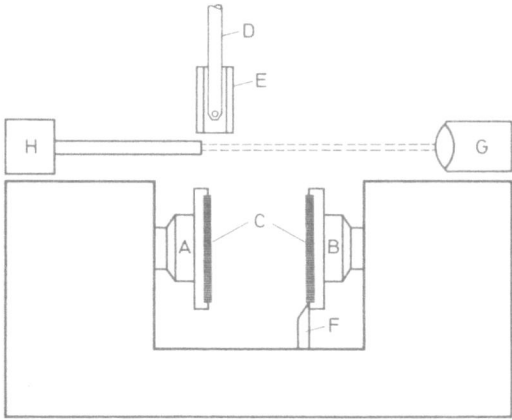

Fig. 12. Apparatus for rapid quenching of materials based on conduction principle

extruded by applying gas pressure at the top of D. Quenching is done by squeezing material between the fixed anvil A and the piston B which is pressurized via a high pressure gas cylinder but held back by a solenoid actuated pin F. After material is extruded it intercepts a beam of light between source G and a phototube H and then by means of a relay, the solenoid releases B. Although the device has, to the author's knowledge, never been used in polymer studies, it offers a convenient means to obtain quenched continuous films of polymer about 3 cm in diameter. Depending upon piston velocity, melt viscosity, etc., it is conceivable that films as thin as a few thousand angstroms and uniform thicknesses up to 100 μ can be produced in this device. Without further preparation such film specimens should then be suitable for dynamic mechanical measurements, electrical resistivity and dielectric studies, for electron microscopy, birefringence investigations, etc. The average rate of cooling in the piston and anvil technique can be estimated by knowing the final film thickness of the quenched specimen, the time it takes for the molten globule on the piston to contact the anvil and the temperature differential of the liquid and the contacting surfaces. For contact times of about 1 msec and a temperature drop of 200° C, an average rate of cooling of 200000° C/sec can be realized in the absence of a thermal gradient within the sample (for very thin film specimens).

While extremely rapid cooling from the liquid state has become a reasonably well established metallurgical method in forming new phases in alloys, it is, in the absence of data on polymeric materials, difficult

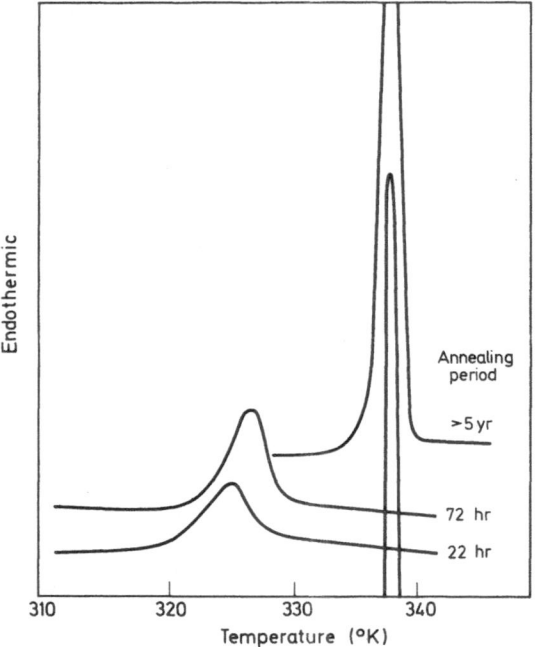

Fig. 13. DSC scans of Aroclor 5460 annealed at 295 °K. Heating rate of 5 °K/min. [Ref. (*105*)]

to assess its importance in the macromolecular sciences. However, since we do not yet understand equilibrium phases in polymers and moreover because the establishment of equilibrium in the solid state of many polymers may take anywhere from a few decades to a few hundred years, the possibility of producing completely new non-equilibrium states, "zero equilibrium reference states" in the condensed phase of polymers is somehow appealing, regardless of any practical implications.

b) Annealing. During rapid processing (i.e., injection molding, extrusion, etc.) polymers are generally subjected to a combination of unfavorable thermal and mechanical excursions, resulting in parts which may contain residual localized stress concentrations leading to cracking or structures which may exhibit undesirable anisotropic behavior. In order to overcome such defects, it is customary to heat treat at temperatures at which stresses or defects are "annealed out" without allowing dimensional changes or deformations to take place. Quite

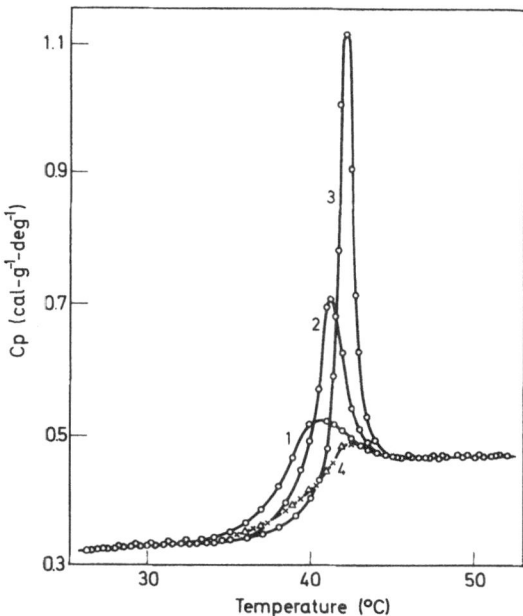

Fig. 14. Changes in specific heat C_p of polyvinyl acetate after annealing for various times at $21°$ C (1—17 h, 2—48 h, 3—7 days, 4— not annealed). Runs were made at a heating rate of $0.5°$/min

differently, the scientific objective of annealing studies is generally to achieve equilibrium in a material or to investigate the kinetic and thermodynamic aspects leading to an equilibrium state. The effects of annealing processes on the thermal behavior of polymers has only relatively recently been recognized and some isolated studies involving energy-volume (*104*—*109*, *115*) dynamic mechanical (*110*, *111*) and dielectric (*112*, *113*) relaxations in polymers as a function of thermal history have been reported. Perhaps the most dramatic demonstration of the effect of annealing on the thermal behavior of an organic glass is given in a paper by Petrie (*105*) for Aroclor 5460, a chlorimated terphenyl.

The results in Fig. 13 illustrate the effects of long annealing periods on the magnitude and slope of the glass transition endotherm. X-ray diffraction studies on this material after annealing did not reveal any crystallinity, thus eliminating the possibility of structural effects. Based on calculations of the energy absorbed in the transition interval, it was

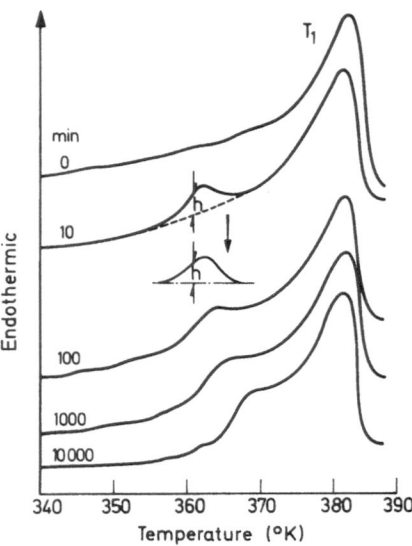

Fig. 15. Fusion curves of annealed polyethylene, unextracted. The annealing temperature is 88° C. The annealing time is indicated on each curve. The procedure for identification of the secondary peak position is also shown.
[Ref. (*115*)]

suggested that the glass is still substantially away from a so-called "equilibrium glassy state" after this long annealing period. In another study by Volkenstein and Sharonov (*109*), the effect of sample preparation on the specific heat of polyvinylacetate was demonstrated (Fig. 14). Drastic differences in the shape of $C_p - T$ curves were observed when quenched samples were annealed at 21° C for various times up to 7 days.

A complete assessment of experimental variables in the study of annealing processes of either polymeric glasses or crystals requires a knowledge of energy-volume relationships and the extent to which relaxations of energy and volume takes place. Specifically, the following information should be established:

1. The enthalpy and volume-temperature relationships and their variations with heating rate (from specific heat and thermal expansion measurement at selected heating rates or under isothermal conditions).

2. The enthalpy and volume of the crystal or glass under equilibrium"[4] conditions at specified annealing temperatures.

---

[4] As used here equilibrium is not a strict thermodynamic term.

3. The magnitude of enthalpy of samples in transition (i.e., melt or glass) zones. (For a given annealing temperature, the maximum energy absorption in the transition interval would correspond to that of an equilibrium state.)

4. Enthalpy and volume changes as a function of annealing time for samples with specified annealing temperatures.

Detailed analytical expressions for energy and volume relaxations of annealing processes occurring in polymeric glasses can be found in papers by Petrie (105) and Kovacs (114) respectively.

On the basis of such experiments over a limited time and temperature range, the thermal behavior of a quenched glass as it approaches its equilibrium state can be quantitatively analyzed, and long relaxation processes (i.e., annealing at low temperatures) can be explored via the familiar time-temperature superposition principles.

While many studies of the effects of annealing on the morphology of crystalline polymers have been reported (116), only a relatively few studies have been devoted to the thermal behavior of annealed polymer crystals (106, 115—119). Wunderlich, Cormier, Keller, and Machin (106) have shown that long term annealing of stirrer crystallized polyethylene increases the amount of superheatable polymer crystals and in addition results in an increase in the zero-entropy melting point of this material. Okamoto (115) has studied the change in a secondary melting peak in low density polyethylene as a function of annealing time up to 120 days (Fig. 15) and concluded that, in contrast to isothermal secondary crystallization which apparently is a diffusion controlled process, the shift of this melting peak toward higher temperatures is controlled by the degree of supercooling.

C. The Glass Transitions
Models, Theories, and Empirical Approaches

A difficulty in the understanding of theories relating to glasses lies in the definition of a glass, or better the lack of a generally accepted one. Vaguely, noncrystalline solids or glasses have been categorized by means of at least two common and negative properties: the absence of long range periodicity as observed by X-ray techniques (amorphicity), and the presence of viscoelasticity (relaxations). Glasses, polymeric or otherwise can be considered as supercooled liquids. But since our concept of liquids are rather vague, it is not surprising that theories about glasses have to date not achieved the formalism of crystal or gas theories. Further limitations to theoretical approaches of glass forming materials are the lack of "realisitic" models[5] capable of linking often well characterized chemical structures of glasses to their physical makeup and the paradox of describing an "ideal" glass. Last but not least there is, in the author's opinion, as yet a communication problem between relaxationists, thermodynamists, morphologists, and chemists whose rather isolated views have contributed much to the confusion which still exists on the nature of noncrystalline polymers and their transitions. Nevertheless, existing glass theories, models and empirical formulations have been useful in explaining the thermal properties of glassy polymers, and in the following discussion a brief account of various approaches will be given.

Detailed overviews on glass transition phenomena of polymers including discussions of theoretical aspects have been published by Kovacs (114), Shen and Eisenberg (120), Kanig (121), Boyer (122) and Ferry (123),

1. Volume Theories

A comprehensive analysis of the "free volume" theories of glass transitions of high polymers has recently been published by Kanig (121).

---

[5] In the past, models that could have originated in toy boxes (i.e., strings, balls, beads, spheres, etc.) or configurations out of car repair shops (i.e., crankshafts, springs, coils, screws, dash pots, bricks, spaghetti, etc.) have been employed.

Over the years considerable controversy has arisen over the definition of free volume. Even in the more general treatments by Enskog (124), Eyring (125) and Frenkel (126) the free volume is operationally not well defined and their theories are not yet complete. Depending on what definition is used (hole-, excess-, vacancies-, cell-, reduced free-, configurational-, expansion or fluctuation-volume), the result differ by as much as two orders of magnitude (121). Although conceptually crude, free volume theories have the advantage over others of being physically plausible and analytically simple to apply.

The hole theory of liquids — based on the conception that within a liquid there are statistically distributed vacancies or holes, whose number and volumia decrease with decreasing temperature has been applied by Fox and Flory (127), Hirai and Eyring (128), Kanig (121), Kovacs (129), Someynsky and Simha (130), and others, to explain various aspects of the glass transition in polymers. In its simplest form the theory defines the molar V and occupied volume $V_o (= N_o \gamma_o)$ as they are related to the number of holes (N) in the following way:

$$V = N_o \gamma_o \left[1 + \exp \{ (\varepsilon_h + p \gamma_h)/R T \} \right] \tag{1}$$

where ε_h and γ_h is the energy and volume of holes.

The thermal expansion coefficient α_h, compressibility β_h and heat capacity C_p of holes can be represented in the following manner:

$$\alpha_h = (\varepsilon_h/R T^2) \exp \{ - (\varepsilon_h + p \gamma_h)/R T \} , \tag{2}$$

$$\beta_h = (\gamma_h/R T) \exp \{ - (\varepsilon_h + p \gamma_h)/R T \} , \tag{3}$$

$$C_{ph} = \eta R (\varepsilon_h/R T) \exp \{ (- \varepsilon_h + p \gamma_h)/R T \} . \tag{4}$$

The free volume at T_g reaches a universal value, which according to Fox and Flory (127) has been estimated to be about 0.02, and based on viscosity measurements by Williams, Landel and Ferry (131) is equal to about 0.025. Thus from Eq. (1), $\varepsilon_h/R T_g \approx - \ln (0.02) = 4$, a value which is useful in estimating ε_h. Assuming that at T_g the holes are "as good as" frozen in, α_h, β_h and C_{ph} are equal to the difference of their respective values just above and below T_g; that is $\Delta \alpha$, $\Delta \beta$ and ΔC_p, quantities which are experimentally accessible. The hole volume, V_h, can be obtained for example by the pressure dependence of shear viscosity,

$$\eta_p = \eta_o \, exp\,(p\gamma_h/RT) \qquad (5)$$

where η_o and η_p are the viscosities at 1 and p atm., respectively.

The universality of the iso-free volume (v_f) at T_g has been debated for several years. According to Kanig, the introduction of a geometric factor (a) which adjusts for structural characteristics of the mixing partners, viz the molecules and holes, yields more realistic results for numerous and widely different glasses. For the system polymer chains holes, the author suggests a value of 3.15 for a $(= V^1 f/V_f)$. It is, however, interesting to note that the *average* specific volume $(\bar{V}_f)$ based on this factor when calculated in the usual manner from thermodynamic qualities $(\Delta\alpha, \Delta C_p, T_g, E_g)$ yields a value of 0.0235, in relatively good agreement with that of Fox and Flory and William Landel and Ferry.

In 1962 Simha and Boyer (*132*) proposed another relationship for calculating the free volume of polymers at T_g. The authors found empirically that

$$(\alpha_l - \alpha_g)\,T \approx k \approx 0.113\,, \qquad (6)$$

where α_l and α_g are the slopes of the volume-temperature line which intersect at T_g. In a recent publication Simha and West and Boyer and Shima (*133*) have concluded that the correlation is still "tolerably well obeyed", considering the experimental uncertainties. Nevertheless, the relationship has been challenged in a paper by Sharma, Mandelkern and Strehling, who concluded that the deviations of k for numerous polymers are intolerable $(0.04 < \Delta\alpha T_g < 0.16)$. It should be mentioned, however, that arguments for or against the rule depend how one defines α [see Ref. (*133*)] and also on how one selects (and plots) the literature data. Kanig (*121*), who selected 11 different polymers concluded that k should not be treated as a constant but as a function of the free volume at T_g. A similar argument was also presented in earlier findings by Miller (*135*) which indicate that it is an iso-free viscosity rather than iso-free volume state that seems to predominate at T_g. More arguments will be presented in a later section of this review.

Other simple free volume models have been developed by Turnbull and Cohen (*136*) and others (*137—139*) to account for the important diffusive transport properties in liquids and glasses and for the liquid-glass transition. In a recent publication (*140*) Turnbull and Cohen have reconsidered the Enskrog theory (*124*) (i.e., developed for dense gases and rigid spheres) and conclude that "at least" molecular transport during liquid-glass transitions can be accounted for within the framework of the classical Van der Waals theory for liquids, provided corrections are made for variations of the "hard core radius" of molecules with temperature.

While Shen and Eisenberg (*12*) have correctly pointed out that there is no real conflict between various free volume theories of glasses, a generally accepted definition of free volume is still lacking. In light of the statistical nature of glasses, especially phase boundaries, the complex basis of inter- and intramolecular interactions, as well as the topology of molecular arrangements of chain molecules, a rigorous description of free volume at this time seems impossible.

2. Thermodynamic Theories

While the "hole" volume model in the Eyring treatment requires no thermodynamic imput, volume is of course very much a thermodynamic entity. In 1958, Gibbs and DiMarzio (*141*) published their approach to the problem of treating polymeric glasses in terms of PVT variables and arrived at the conclusion that the glass transition is "in fact, the experimental manifestation of a second-order transition" T_2 in the Ehrenfest sense (*142*)[6]. In recognizing the shortcomings of earlier pure kinetic theories in explaining the configurational and conformational irregularities of chain molecules in the supercooled state and their failure to explain the kinetic phenomena themselves, these investigators viewed the cooling of glass forming materials strictly as an entropic process, which, when carried out slowly and long enough, eventually leads to an equilibrium glassy state at T_2 with the usual zero configurational entropy production. The relaxation behavior governing T_g, viz. kinetic effects, were explained as a consequence of the very small values in the number of configurations available to the system in the region (ΔS conf. $\to 0$) about T_2. In an extention of the $G-D$ theory, Adam and Gibbs (*143*) recently attempted to find a quantitative relation between the observable T_g and hypothetical T_2 and thereby link the kinetic features of the glass transition to the underlying fundamentally important thermodynamic manifestations. The results of this study showed that T_g/T_2 was on the

---

[6] An interesting discussion of the development of the Gibbs and DiMarzio equilibrium theory as related to earlier treatments by Lewis and Gibson, Kaufmann, Simon, Davis and Jones and kinetic theories is given in Ref. (*120*).

average 1.30 and did not deviate much for a large number of different glasses. These findings were in agreement with earlier experimental results by Bestul and Chang (144), and Passaglia and Kevorkian (145). Moreover, in recognizing that the transition probability of a "co-operatively rearranging region" (upon which as least part of the theory rests) is related to the reciprocal of the relaxation time in kinetic theories, Adam and Gibbs by means of a temperature shift factor (α_T) arrived at a formula:

$$- \log \alpha_T = [a_1 (T - T_s)] [a_2 + (T - T_s)] \tag{7}$$

of the WLF (William, Landel and Ferry) type in which the "universal but not always right" Williams parameters C_1 and C_2 are replaced by a_1 and a_2. The important difference between the $A - D$ and WLF parameters is that

$$a_1 = 2.303 \, \Delta \mu S_c^* / k \Delta C_p T_s \ln (T_s / T_2) \,, \tag{8}$$

where T_s is a reference temperature and $\Delta \mu S^* / \Delta C_p$ is a Gibbs free energy barrier for molecular segment, and

$$a_2 = \frac{T_s \ln (T_s / T_2)}{1 + \ln (T_s / T_2)} \,, \tag{9}$$

Thus the validity of these parameters depend essentially on the constancy of $\Delta \mu S_c^* / \Delta C_p$, which as the authors believe, "can be understood at least qualitatively from a molecular point of view."

At this point ($\sim$ 1965) in the development of Tg theories, both the relaxationists and thermodynamists could at last have celebrated an unification of thought. However, the old question whether the formalism of phase transformations in a thermodynamic sense applies to polymeric glasses came up again. In the above molecular-kinetic theory, the single most important prerequisite is the existence of a T_2, a transition point of the second kind, in which quantities such as specific heat, thermal expansion coefficient, isothermal elastic coefficient and other compliance coefficients related to the statistico-mechanical treatment of Tisza (146) has a singularity rather than a discontinuity. The differentiation between a first and second order phenomena can generally readily be made from calorimetric measurements.

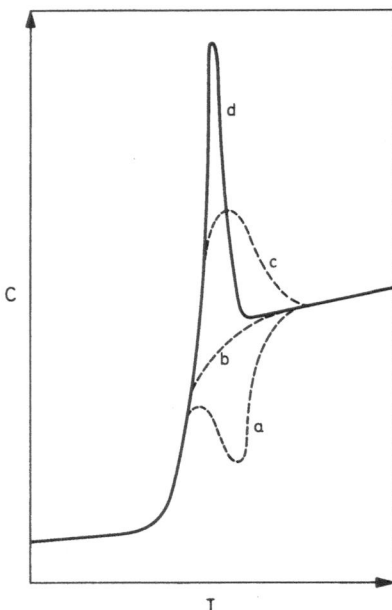

Fig. 16. Specific heat c as a function of temperature for a typical glass. a quenched sample, b slowly cooled sample, c long time annealed sample, d infinitely long annealed sample. [Ref. (*147*)]

Thus the shape of $C_p - T$ curves such as in case *a*, *c* and *d* in Fig. 16 suggest a heat of transition (i.e., 1st order phenomena) while curve *b* implies a 2nd order type of transition. The shape of such curves (see also Fig. 13 and 14) has in the past misled investigators and resulted in misconceptions and misinterpretations of the nature of glass transitions in polymers (*78, 147, 148, 149*). Ueberreiter based on such evidence in earlier work, proposed a first order mechanism with an overlapping associated freezing process. In a recent detailed review on the thermal behavior of annealed organic glasses (*105*) it was demonstrated, however, that it is not necessary to postulate a first order process to account for the abrupt increase in enthalpy at the glass transition, a conclusion which was also reached very recently by Bruns, Melidorn and Ueberreiter (*150*).

All in all, at this time a first order thermodynamic glass transition can definitely be excluded. However, the controversy over the existence of a second order transition (even though a hypothetical one) in the

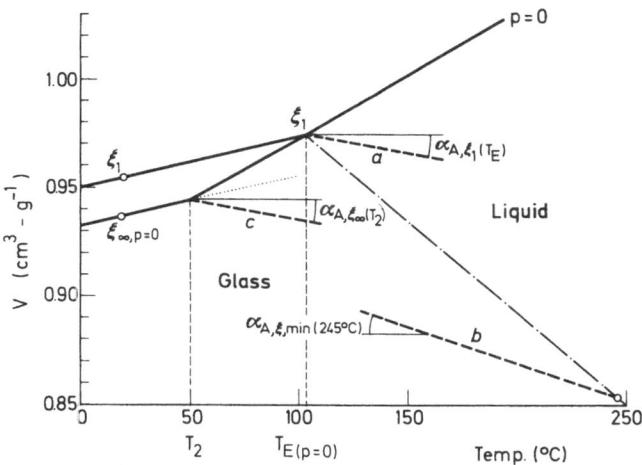

Fig. 17. Volume temperature diagram illustrating the freezing process in polystyrene from the melt to the glassy state. [Ref. (151)] (Symbols: ξ_1 configurational parameter under normal conditions of pressure and cooling rate); ξ_∞ configurational parameter at equilibrium; $T_2 = \lim_{T \to 0} T_E$; α_A specific volume expansion coefficient of the liquid approaching equilibrium T_E temperature of initial freezing

Ehrenfest sense is not yet resolved. According to the $G-D$ and $A-G$ theories, the "freeze in" temperature reaches a limit with decreasing cooling rate at T_2 at which point the material reaches its "best possible" configuration. This configurational limit (ξ_∞) corresponds to a maximum packing density for a given glass which as noted by Breuer and Rehage (151) at normal pressures can experimentally never be achieved. However, at high enough pressures and temperatures such a material could approach an equilibrium state rather closely. In an attempt to demonstrate this, Breuer and Rehage examined the dependence of volume on temperature, pressure, and time in atactic polystyrene (the accepted standard for the experimental verification of glass theories). Their results are schematically summarized in Fig. 17.

At normal pressures ($P = 0$) and finite cooling rates, one obtains for constant internal equilibrium (A) the volume-temperature relation (curve a) according to the equation:

$$\delta\varepsilon, A = \left(\frac{\partial V}{\partial T}\right)_{\xi, A} = \alpha_A - \chi_A \left(\frac{\partial P}{\partial T}\right)_{\xi, A} = \alpha_A - \chi_A \left(\frac{\Delta\alpha}{\Delta\chi}\right). \tag{6}$$

Similarly, at high pressures and temperatures ($\sim 245°$), $\Delta\alpha = \alpha_A$ and $\alpha\,\xi, A$ reaches a minimum, such that

$$\alpha\xi_{A,\,min} = \alpha_A(1 - \chi_A/\Delta\chi)\,. \tag{7}$$

Which leads to curve b.

At normal pressures the freezing process to the glassy state shifts toward lower temperatures with decreasing cooling rate. Therefore, at infinitesimal small rates (curve c),

$$\left(\frac{d\xi}{dT}\right)_E = 0 \quad \text{for} \quad (\dot{T} \to 0) \tag{8}$$

as required in the Ehrenfest process, one can except a more perfect configuration (ξ_∞) of the glass and would hope to find a T_2, which according to Gibbs and DiMarzio represents a 2nd order transition.

However, the slopes of curves c and a are practically identical (i.e., a T_2 could not be found) and all three curves, a, b and c, deviate strongly from the "freeze in" curve d. Based on these data and supporting evidence from volume retardation analysis, Brewer and Rehage (151) have concluded that a 2nd order transition within the limiting cases of high temperatures and pressures and infinitely slow cooling rates at normal pressures does not exist.

The authors argue convincingly that an analytical description of the glass transition region requires, in addition to two thermodynamic variables (such as pressure and temperature), also an "order" variable, which in the case of polysyrene involves more than one independent configurational parameter.

3. Kinetic Theories

The kinetics of ordering in solid solutions and even more so rate processes during phase transformations are both expermentally and theoretically very difficult subjects. Furthermore, in light of the previous discussion it is highly questionable whether order-disorder processes as they occur in simple glass forming (i.e., metal alloys) substances also take place in polymers. Nevertheless, attempts are being made to improve earlier kinetic theories (124—126) by taking into account systems of more complex molecular dynamics in which the free volume is less clearly defined.

Briefly, the rate of molecular rearrangements in the condensed noncrystalline (dense fluids) phase takes place presumably by the elementary processes of shear viscosity and diffusion, which in combination consist of cooperative motions of molecular assemblies around vacancies or holes. For a first approximation one assumes a first order kinetics; thus, the rate of decrease of the number of vacancies or holes at constant pressure is according to Hirai and Eyring (128).

$$\left(\frac{\partial N_h}{\partial t}\right)_p = \frac{1}{\tau}\{N_h(T) - N_{h\infty}\} \tag{9}$$

or

$$\left(\frac{\partial v}{\partial t}\right)_p = \frac{1}{\tau}(v - v_\infty) \tag{10}$$

where $N_{h\infty}$ and v_∞ are number and volume of holes of the system at infinitely long times and τ_β is the relaxation time (i.e., the equivalent of a rate constant) for bulk contraction at constant p, defined as

$$\tau_\beta = \frac{h}{kT}\exp\left\{\varepsilon_j - T\ln\left[\frac{F^*(B)}{F_h} - 1/2\,p\gamma_h\right]\middle/RT\right\}. \tag{11}$$

Relationships between the quantities given in Eq. 11 are shown in Fig. 18. The use of Eq. (11) leads to the following relationship between the rate constant (τ_β) and bulk viscosity (related to the shear viscosity)

$$\eta_B = \tau_\beta k_h = \frac{h}{v_h}\exp\{[\varepsilon_h + \varepsilon_j + 1/2\,pv_h - T\ln F^*/F_h]/RT\} \tag{12}$$

and ultimately to the "activated state" relation between viscosity and self diffusivity.

$$\eta D = n^{1/3}kT. \tag{13}$$

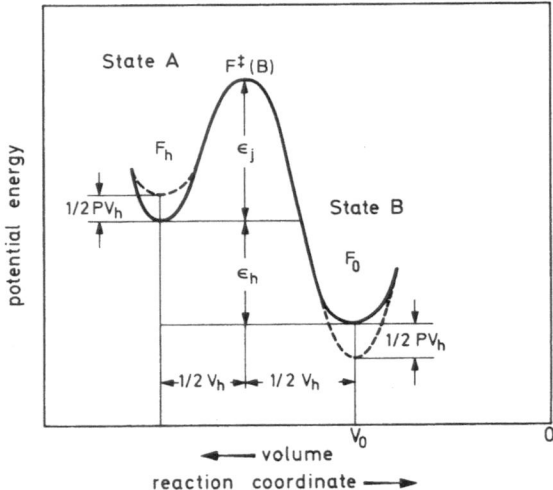

Fig. 18. Reaction coordinate and potential energy diagram for bulk flow

An improved treatment for D has recently been given by Turnbull and Cohen (140), by taking into consideration the variable magnitude of the diffusive displacement in the Enskrog theory. Their analysis shows that the diffusion coefficient in an Enskrog fluid is falling precipitously with increasing density and that this density trend (which was attributed to back scattering) eventually leads to continuous solidification as the glass transition is approached. These investigators conclude, as already noted in a previous section, that the molecular transport manifestation of a liquid-glass transition can be deduced entirely according to the Van der Waal's physical model of liquids provided corrections in the Enskrog fluid are made for the effects of back scattering, viz softness of real glasses.

Modification of the Hirai-Eyring rate theory have recently been made by Wunderlich, Bodily and Kaplan (152) for cases of continued cooling and heating. By dividing Eq.(9) by a heating or cooling rate, $q, (= dT/dt)$ one obtains

$$\left(\frac{\partial N}{\partial T}\right)_p = \frac{1}{q \cdot \tau}(N_h T) - N_{h\infty} .$$

(14)

The solution of Eq. (14) is

$$N = N_a \exp - \varphi(T) + \exp - \varphi T \int_{T_a}^{T} \frac{N^*(T')}{q\tau} \exp \varphi(T') \, dT' \qquad (15)$$

where N_a and T_a indicate the number of high energy conformations and the temperature at zero time and

$$\varphi(T) = \int_{T_a}^{T} \frac{dT}{q\tau}. \qquad (16)$$

It is to be noted that N_a is quite differently from $N_h(T)$ in Eq. (9), if the glass attains more than its equilibrium number of conformations. Here N_a is a parameter to describe various frozen in states on cooling a liquid through its glass transition interval.

Other useful recent contributions to the understanding of the kinetic feature in glass transitions can be found in papers by Miller (153), Anderson (154), Kovacs (114) and Volkenstein and Saranov (109).

4. Magnitude of the Glass Transition

a) **Free Volume, Thermal Expansion and Molecular Heat.** Table 2 lists experimental values of the important thermal properties in the glass transition interval of 54 different polymers. The only missing property for a complete thermophysical description is the change in compressibility, $\Delta\beta$, which so far has been determined for only a few samples.

Polymers have been grouped into the following 10 categories: aliphatic polyolefins, polystyrenes, polyvinyl halides, polyvinyl esters, cellulose esters, polyvinyl ethers, poly-ethers, aromatic polymers, copolymers and miscellaneous polymers. The molar cohesive energies (E_g) were calculated from their respective group increments according to the following equation:

$$E_g = \Sigma\varepsilon_i + \Sigma\varepsilon_i \, 1.73 \, \frac{(\Sigma v_i - v_g)}{v_g} \qquad (17)$$

where $\Sigma \varepsilon_i$ and Σv_i is the sum of group molar cohesive energies based on a method described by Bunn (162). For some polymers experimental values for ΔC_p were not accessible. In those circumstances the rule of constant heat capacity increment (161) was applied and ΔC_p's were calculated based on the repeat unit structure of the polymers. Unfortunately, for cellulose esters, v_g and for polyoxides, $\Delta \alpha$, could not be found in the literature, and computation of the remaining quantities could not be made for these systems. The quantities ε_h, v_h, v_o, n_o and n_h (Table 2b) are defined in Section 1b under free volume theories. For uniformity these parameters were calculated based on the equations given in Kanig's paper, in which also an implicit equation for the specific free volume φ, and a geometric factor a, are derived.

While the primary objective of compiling the data in Table 2 was to assess thermodynamic parameters for polymeric glasses which serve to illustrate order of magnitude changes in the glass transition interval, the values were also used to test various theoretical and empirical relationships. As already noted by Kanig for a smaller number of polymers, φ_1, the specific free volume at T_g calculated from the thermodynamic quantities T_g, $\Delta \alpha^*$, ΔC_p and E_g is not a constant but varies for different polymers. Considerable scatter was also observed in the values for a, $\Delta \alpha^* T_g$ and v_o/v_h. Thus, when a large body of structurally different polymers are considered and the results from many laboratories are utilized indiscriminately, an isofree-volume state in polymeric glasses does not seem to exist.

An attempt was also made to correlate the hole energy, ε_h, with glass transition temperature. In the expression developed by Harai and Eyring, the free volume, $N_h v_h$ is related to the hole energy by a Bolzmann type distribution function:

$$\frac{N_h v_h}{N_o v_o} = e^{-\varepsilon_h/RT} . \tag{18}$$

Now, by using the calculated free volume φ_1^* at T_g (rather than assuming a constant value) we obtained according to Kanig (123)

$$\varepsilon_h = -R T_g [\ln \varphi_1^* + (1 - \varphi_1^*)] . \tag{19}$$

In Fig. 19 the hole energy calculated from the data in Table 2b is plotted as a function of T_g. Considering the large number of different

Table 2a. Thermodynamic properties of polymeric glasses

Polymer	No.	Ref.	M_o	T_g (°K)	$\alpha_l \times 10^4$ (deg^{-1})	$\alpha_g \times 10^4$ (deg^{-1})	$\Delta\alpha \times 10^4$ (deg^{-1})	v_g (cm^3·g^{-1})	$\dfrac{\Delta\alpha^*}{\Delta\alpha}\,v_g$	ΔC_p (cal-mole^{-1})(deg^{-1})
I. *Polyolefins*										
Polyethylene	1	134 (7)	28	140	5.31	2.01	3.30	1.035	3.14	—
		45		237	—	—	—	—	—	2.37
	2	165		—	5.50	3.00	2.50	0.969	2.58	(2.70)
Polypropylene (Compl. amorphous)	3	121 (d)	42	243.5	—	—	3.72	1.129	4.20	3.806 (?)
		121 (e)		260						
	4	134 (29)		258	9.40	2.20	7.20	1.130	6.37	—
	5	134 (0)		250	8.10	4.40	3.70	1.130	3.28	—
		134 (30)		255	—	—	4.00	—	—	6.15
		159		259	—	—	—	—	—	(5.4)
Polyisobutylene (PIB)	6	121 (b)	56	199	6.18	1.48	4.70	1.070	4.48	—
	7	134 (18)		202	6.00	1.50	4.50	1.072	4.22	—
	8	134 (0)		205	5.40	1.35	4.05	1.078	3.75	—
		121 (f)		—	—	—	—	—	—	5.32
		121 (h)		—	—	—	—	—	—	5.325
		45		—	—	—	—	—	—	5.32
Polybutadiene (PBD)	9	134 (19)	54	188	7.80	2.00	5.80	—	5.66	—
		121 (f)		190	—	—	—	1.025	—	8.01
Polyisoprene (PIP)	10	121	68	201	6.16	2.06	4.10	1.038 (f)	3.956	7.35 (f)
		121 (j)		199	—	—	—	—	—	7.42
		121 (h)		200	—	—	—	—	—	7.48
		45		199	—	—	—	—	—	10
(N.R.)	11		68	201	6.02	2.00	4.02	1.040	3.75	—
				159	—	—	—	—	—	8.3

Table 2a (continued)

Polymer	No.	Ref.	M_o	T_g (°K)	$\alpha_l \times 10^4$ (deg^{-1})	$\alpha_g \times 10^4$ (deg^{-1})	$\Delta\alpha \times 10^4$ (deg^{-1})	v_g (cm^3·g^{-1})	$\dfrac{\Delta\alpha^*}{\Delta\alpha}\,v_g$	ΔC_p (cal-mole^{-1}) (deg^{-1})
Poly(4-Me-1-pentene) (PMP)	12	134 (23)	85	302	7.61	3.83	3.78	1.196	3.16	—
		155		—	—	—	—	—	—	8.1
II. Polystyrenes										
Polystyrene (PS)	13	121	104	368	5.50 (a)	1.80	3.70	0.973	3.74	6.35 (c)
	14	121 (b)		373	5.5	2.5	3.0	0.97	3.08	—
		156		373	—	—	—	—	—	9.50 (?)
		157		373	—	—	—	—	—	7.30
		114		375	—	—	—	—	—	9.36 (?)
	15	121 (l)		370	4.44	1.75	2.69	0.966	2.80	6.55
		159		373	6.00	2.60	3.40	—	—	—
Poly(α-methylstyrene) (PMS)	16	134 (10)	119	453	5.40	2.40	3.00	0.976	3.07	(7.50)
Poly(p-chlorostyrene) (PCS)	17	121 (l)	135.5	383	3.79	1.30	2.48	0.8564	2.9	7.48
Poly(p-bromostyrene) (PBS)	18	121 (l)	233	391	3.19	0.92	2.27	0.6706	3.4	8.77
Polystyrene sulfonic acid	18a	159	169	402	6.20	2.10	4.10	—	—	—
III. Polyvinylhalides										
Polyvinylchloride (PVC)	19	134 (24)	62.5	355	—	—	2.15	0.733	2.93	4.25 (f)
		121 (b)		355	5.2	2.1	3.10	0.75 (f)	4.30	—
	20	121 (g)		352	4.67	1.40	3.25	0.723	4.5	4.44
	21	121 (e)		343	4.20	1.84	2.36	0.728	3.24	4.13
		158		—	—	—	—	—	—	4.40
					2.69	1.25	1.44	0.4856	2.97	(5.4)

Table 2a (continued)

Polymer	No.	Ref.	M_o	T_g (°K)	$\alpha_l \times 10^4$ (deg⁻¹)	$\alpha_g \times 10^4$ (deg⁻¹)	$\Delta\alpha \times 10^4$ (deg⁻¹)	v_g (cm³-g⁻¹)	$\dfrac{\Delta\alpha^*}{\Delta\alpha}\,v_g$	ΔC_p (cal-mole⁻¹)(deg⁻¹)
Poly(trifluorochloroethylene) (PTE)	22	134	117.5	352	2.51	1.49	1.02	0.4856	2.10	(3.98)
Polyvinylidine chloride (PVIC)		166		255	4.78	2.37	2.41	—	4.00	(5.5)
	23	168	135	—	—	—	—	0.605	—	—
IV. *Polyesters*										
Polyvinylacetate (PVA)	24	121 (b)	86	302	5.98	2.07	3.91	0.839	4.65	—
		134 (22)		305	5.90	2.30	3.60	0.844	4.26	—
	25	121 (k)		299	—	—	—	—	4.50 (h)	8.61
Poly(methylmethacrylate) (PMMA) atactic	26	121 (b)	100	378	5.00	—	3.05	—	3.54	7.56 (c)
	27	134 (27)		378	4.60	2.15	2.45	0.870	2.82	7.0 (h)
	28	121 (l)		375	4.82	1.65	3.27	0.821	3.86	7.1
	29	121 (o)		377	5.15	2.07	3.08	0.861	3.58	—
		159		375	—	—	—	—	—	7.0
Isotactic		159		319	—	—	—	—	—	11.1
Syndiotactic		159		394	—	—	—	—	—	7.6
Atactic		158		—	—	—	—	—	—	8.2
Poly(t-butyl acrylate) (PBA)	30	121 (l)	132	304	6.20	1.57	4.63	0.9809	4.72	9.87
Poly(methylacrylate) (PMAC)	44	134 (20)	86	282	5.60	2.70	2.90	0.807	3.59	(5.4)
	45									(7.1)
Poly(ethylene terephthalate) (PET)	46	134 (8)	182	337	4.50	1.80	2.70	0.771	3.51	15.51
		163		—	3.70	1.60	2.10	—	2.84	—
	47	161		342	—	—	—	0.77	—	15.60

Table 2a (continued)

Polymer	No.	Ref.	M_o	T_g (°K)	$\alpha_l \times 10^4$ (deg⁻¹)	$\alpha_g \times 10^4$ (deg⁻¹)	$\Delta\alpha \times 10^4$ (deg⁻¹)	v_g (cm³·g⁻¹)	$\Delta\alpha^*$ $\frac{\Delta\alpha}{v_g} v_g$	ΔC_p (cal-mole⁻¹)(deg⁻¹)
IV. *Polyesters (continued)*										
Poly(ethylmethacrylate) (PEMA)	31	134 (27)	114	338	5.40	2.75	2.65	0.900	2.95	(7.1)
Poly(propylmethacrylate) (PPMA)	32	134 (27)	128	308	5.80	3.15	2.65	0.932	2.85	(7.1)
Poly(n-butylmethacrylate) (PBMA)	33	134 (27)	142	293	6.10	3.80	2.30	0.948	2.30	(7.1)
Poly(n-hexylmethacrylate) (PHMA)	34	134 (27)	170	268	6.80	4.40	2.40	0.972	2.50	(7.1)
Poly(n-octylmethacrylate) (POMA)	35	134 (27)	198	253	6.0	4.15	2.35	1.002	2.35	(7.1)
Poly(n-dodecylmethacrylate) (PDMA)	36	134 (27)	254	208	6.80	3.80	3.00	1.016	2.96	(7.1)
Poly(piperlide acrylate) (PPA)	42	134 (32)	156	381	4.50	2.00	2.50	1.047	2.39	(5.4)
Poly(morpholideacrylate) (PMA)	43	134 (32)	158	418	4.40	2.00	2.40	0.861	2.79	(5.4)
V. *Polyvinyl ethers*										
Polyvinylmethyl ether (*PVME*)	37	134 (41)	58	251	6.45	2.16	4.29	0.9327	4.60	(5.4)
Polyvinylethyl ether (*PVEE*)	38	134 (41)	72	240	7.26	3.03	4.23	1.006	4.21	(5.4)
Polyvinylisopropyl ether (*PVIE*)	39	134 (41)	86	261	6.69	3.06	3.63	1.056	3.43	(5.4)
Polyvinyln-butyl ether (*PVBE*)	40	134 (41)	100	217	7.26	3.90	3.36	1.0163	3.30	(5.4)
Polyvinyln-hexyl ether (*PVHE*)	41	134 (41)	128	199	6.66	3.75	2.91	1.014	2.87	(5.4)
VI. *Aromatic polymers*										
Poly(dimethylphenylene ether) (PDMPO)	48	134 (o)	121	480	5.13	2.04	3.09	0.970	3.20	—
		45		—	—	—	—	—	—	6.90
Polyphenylene ether (PPO)	49	100	92	363	2.48	0.63	1.90	0.78	2.45	5.60
Polyphenylphenylene ether (PPPO)	50	102	246	493	—	—	1.70	0.86	2.00	16.50

Table 2a (continued)

Polymer	No.	Ref.	M_o	T_g (°K)	$\alpha_l \times 10^4$ (deg⁻¹)	$\alpha_g \times 10^4$ (deg⁻¹)	$\Delta\alpha \times 10^4$ (deg⁻¹)	v_g (cm³·g⁻¹)	$\dfrac{\Delta\alpha^*}{\Delta\alpha}\,v_g$	ΔC_p (cal-mole⁻¹) (deg⁻¹)
Polysulfone (PSO)	51	160	432	460	6.3	3.1	3.2	0.95	3.30	30.80
Polyquinoxaline (PPQ)	52	67	558	571	1.84	0.59	1.25	0.76	1.64	42.70
Polycarbonate (PC)	53	121 (n)	254	416	—	—	3.21	(0.81)	3.96	15.24
	54	178		418	—	—	2.82	(0.81)	3.50	—
Selenium		134 (39)		302	1.11	0.415	0.695	0.242	2.88	—
		121 (h)		305			6.53	—	2.70	2.13
		121 (f)		304			—	—	3.51	—
VII. Polyoxides										
Polyoxybutane		45	72	185	—	—	—	—	—	13.5
		165		194	—	—	—	1.00	—	—
Polyoxypropylene		45	58	198	—	—	—	—	—	7.65
		165		—	—	—	—	1.001	—	—
Polyoxymethylene		45	30	200	—	—	—	—	—	(5.4)
				—	—	—	—	0.80	—	—
Polyoxyethylene		165		206	—	—	—	0.88	—	(8.1)
VIII. Copolymers										
Vinylidine chloride-acrylonitrile copolymer (PVCA)	55	168	88.5	317	3.20	0.95	2.15	0.651	3.31	(5.4)
Vinylidine chloride-vinylchloride-copolymer (PVCVC)	56	166	80	268	4.53	2.23	2.30	(0.67)	3.40	(5.4)
Poly(butadiene-styrene) (PBS) Polystyrene – 8.58%	57	158	62.3	193	7.63	2.05	5.58	1.010	5.57	6.89

Table 2a (continued)

Polymer	No.	Ref.	M_o	T_g (°K)	$\alpha_l \times 10^4$ (deg⁻¹)	$\alpha_g \times 10^4$ (deg⁻¹)	$\Delta\alpha \times 10^4$ (deg⁻¹)	v_g (cm³·g⁻¹)	$\Delta\alpha^* \dfrac{\Delta\alpha}{v_g}$	ΔC_p (cal-mole⁻¹)(deg⁻¹)
Polystyrene — 22.61%	58	158	65.5	2.3	7.20	2.15	5.05	1.005	5.04	7.10
Polystyrene — 25.5%	59	158	66.6	212	7.09	2.18	4.91	0.999	4.92	6.95
Polystyrene — 43%	60	158	75.4	237	6.75	2.27	4.48	0.987	4.55	7.37
Polyethylene terephthalate-Sebacate (80—20 mole-ratio)		158		296				0.778		19.94
IX. Miscellaneous polymers										
Poly hexamethylene adipamide (PHMA) (partially crystalline)	61	165	226	322	—	—	—	0.910	—	—
		166		323	4.9	3.9	1.0	—	1.14	(38)
		167		—	—	—	—	—	—	~10
Poly(dimethyl siloxane) (PDS)	62	134(0)	74	150	8.5	4.5	4.0	0.905	4.40	(5.4)
		165		—	—	—	—	1.020	—	—
Polyacrylonitrile (PAN)	63	164	53	360	5.3	2.4	2.9	—	3.44	6.0
		168		360	—	—	—	0.85	—	(5.4)
X. Cellulose esters										
Triacetate		167		445	—	—	—			
Triproprionate		167		406	5.6	3.1	2.5			
Tributyrate		167		365	5.6	3.2	2.4			
Trivalerate		167		289	7.4	3.6	3.8			
Tricaproate		167		223	5.5	3.7	1.8			
Triheptanoate		167		184	7.3	3.9	3.4			
Tridecanoate		167		204	9.9	3.7	6.2			

Table 2b. Thermodynamic properties of polymeric glasses

Polymer No.	ΔC_p	T_g	$\Delta \alpha^*$	E_g	$\varphi_1^* \times 10^2$	a	v_o/v_h	ε_h	n_o/n_h
1	2.37	140	3.14	1360	0.7618	6.01	6.85	1093	20.1
2	2.37	237	2.58	1360	0.8614	7.61	5.08	1792	23.9
3	3.8	244	4.2	3300	2.08	3.06	6.34	1397	7.41
4	3.81	258	6.37	3300	3.96	2.06	4.96	1157	4.88
5	3.81	250	3.28	3300	1.50	3.91	7.19	1589	9.07
6	5.32	199	4.48	4750	1.94	2.73	9.98	1166	5.04
7	5.32	202	4.22	4750	1.81	2.90	10.22	1210	5.29
8	5.32	205	3.75	4750	1.56	3.26	10.83	1288	5.82
9	8.01	189	5.66	4990	2.11	3.19	12.85	1077	3.60
10	7.42	201	3.956	6200	1.59	3.31	14.70	1254	4.18
11	7.42	201	3.75	6200	1.48	3.49	15.18	1283	4.37
12	8.1	302	3.16	5276	1.34	5.60	12.72	1986	5.76
13	6.35	368	3.74	7320	2.91	2.64	8.82	1867	3.76
14	7.3	373	3.08	7320	2.03	3.71	10.78	2153	4.46
15	6.55	370	2.8	7320	1.90	3.61	10.51	2181	4.88
16	7.5	453	3.07	10708	2.97	2.60	10.38	2282	3.14
17	7.48	383	2.9	8540	2.08	3.43	11.51	2192	4.07
18	8.77	391	3.4	9295	2.50	3.19	11.98	2098	3.24
19	4.25	355	2.93	4444	1.88	3.70	6.77	2101	7.67
20	4.44	352	4.5	4444	3.33	2.56	5.62	1695	5.14
21	4.13	343	3.24	4444	2.15	3.24	6.42	1940	7.05
22	3.98	325	2.1	4438	1.16	4.72	8.41	2226	10.0
23	5.5	255	4.0	5650	2.22	2.69	9.41	1428	4.67
24	10.3	305	4.26	6430	1.95	4.41	13.29	1783	3.76
25	8.61	299	4.5	6430	2.40	3.45	11.45	1630	3.54
26	7.56	378	3.54	7820	2.55	3.14	10.31	2016	3.70
27	7.1	378	2.82	7820	1.95	3.60	11.01	2210	4.55
28	7.1	375	3.86	7820	3.00	2.70	9.47	1883	3.41
29	7.1	377	3.58	7820	2.70	2.90	9.80	1968	3.67
30	9.87	304	4.72	9135	3.03	2.62	13.34	1519	2.39
31	7.1	338	2.95	9700	2.15	2.74	12.30	1913	3.68
32	7.1	308	2.85	11450	2.04	2.35	13.75	1773	3.47
33	7.1	293	2.3	12920	1.51	2.54	16.36	1857	3.96
34	7.1	268	2.5	16020	1.66	1.85	16.92	1650	3.48
35	7.1	253	2.35	19470	1.51	1.60	18.51	1607	3.52
36	7.1	208	2.96	24010	1.71	1.01	18.56	1269	3.08
37	5.4	251	4.6	6900	2.99	1.84	9.02	1261	3.58
38	5.4	240	4.21	9550	2.87	1.41	10.16	1225	3.32
39	5.4	261	3.43	11590	2.49	1.42	10.94	1402	3.56
40	5.4	217	3.3	13380	1.98	1.26	12.69	1263	3.89
41	5.4	199	2.87	15360	1.51	1.25	14.66	1263	4.44
42	5.4	381	2.39	13360	2.35	1.80	10.55	2090	3.92
43	5.4	418	2.79	11970	3.10	1.75	9.04	2072	3.45
44	5.4	282	3.59	4180	1.73	4.09	8.61	1715	6.57
45	7.1	282	3.59	4180	1.42	5.48	10.68	1825	6.49
46	15.51	337	3.51	11905	1.87	4.31	21.69	1999	2.41
47	15.6	342	2.84	11905	1.41	5.31	24.42	2216	2.85

Table 2 b (continued)

Polymer No.	ΔC_p	T_g	$\Delta \alpha^*$	E_g	$\varphi_1^* \times 10^2$	a	v_o/v_h	ε_h	n_o/n_h
48	6.9	480	3.2	11750	3.72	2.07	9.38	2212	2.75
49	5.6	363	2.45	7078	1.68	3.59	10.09	2226	5.76
50	16.5	493	2	29645	2.00	3.09	28.25	2861	1.73
51	30.8	460	3.3	25710	2.18	4.33	35.23	2590	1.26
52	42.7	571	4.1	35100	3.19	3.69	36.01	2799	0.84
53	15.24	416	3.96	25130	4.44	1.71	20.68	1778	1.04
54	15.24	418	3.5	25130	3.70	1.93	21.71	1930	1.19
55	5.4	317	3.31	5590	2.04	3.28	8.74	1827	5.48
56	5.4	268	3.4	5045	1.74	3.51	9.59	1626	5.86
57	6.89	193	5.57	5050	2.31	2.72	11.33	1065	3.72
58	7.1	213	5.04	5310	2.20	2.97	11.51	1196	3.85
59	6.95	212	4.92	5400	2.17	2.91	11.54	1196	3.89
60	7.37	237	4.55	5700	2.12	3.2	11.75	1348	3.92
61	38	323	1.14	23800	4.55	16.25	72.87	4553	3.00
62	5.4	150	4.4	3395	1.24	3.94	12.00	1009	6.61
63	6	360	3.44	5130	2.06	3.94	8.25	2069	5.75

Legend to Table 2 a and 2 b:

T_g	Glass transition temperature, $\bar{T}_g$ average value.
α_l	Specific volume expansion coefficient in the liquid state $(\partial v/\partial T)_l$.
χ_g	Specific volume expansion coefficient in the glassy state $(\partial v/\partial T)_g$.
$\Delta \alpha$	$\alpha_l - \alpha_g$.
v_g	Specific volume of polymer at T_g.
$\Delta \alpha^*$	$\Delta \alpha/v_g = (\alpha v/\alpha T)_l - (\alpha v/\alpha T)_g/v_g$, $\bar{\Delta} \alpha$ = average value.
ΔC_p	Change in heat capacity in glass transition interval, $\bar{\Delta G}_p$ is average value.
E_g	Molar cohesive energy as calculated from Ref. (162).
ε_h	Molar hole energy [as calculated in Ref. (121)] $\varepsilon_h = \dfrac{\Delta C_p}{\Delta \alpha^*} \cdot \dfrac{v_h}{v_o} \cdot \varphi_2^*$.
v_o	Volume occupied by molecules.
v_h	Hole volume at the glass transition temperature as defined in Ref. (125).
n_2/n_1	Ratio of monomer units to holes defined in Ref. (125) and calculated in Ref. (121).
φ_1^*	Specific free volume as derived at in Ref. (121).
a	Geometric factor as derived at in Ref. (121).
M_o	Molecular weight of monomeric unit.

polymers covered in this curve with glass transition temperatures ranging from 150—500 °K, the correlation between ε_h and T_g is exceptionally good.

A numerical evaluation of the data shows that $\varepsilon_h/R T_g = 2.88 \pm 0.22$ and it is tempting to consider this value as the "universal constant" for

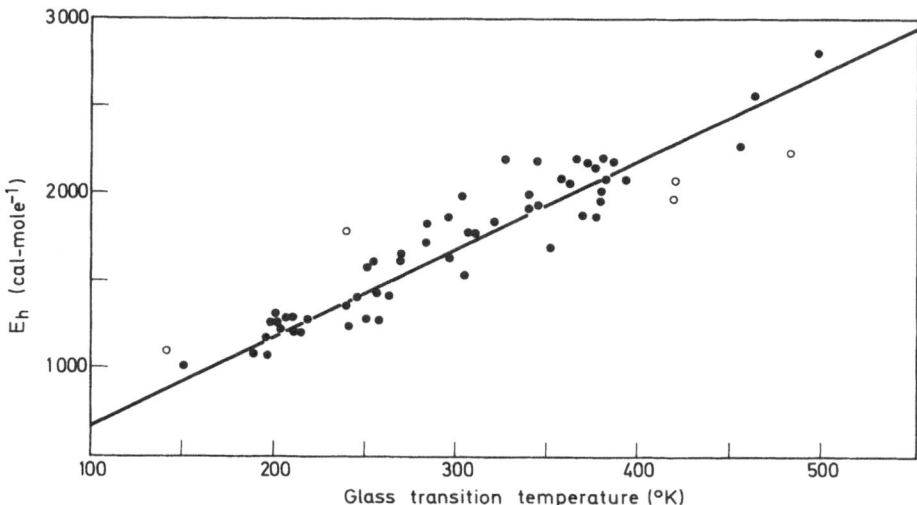

Fig. 19. Plot of hole energy vs. glass transition temperature

polymeric glasses, particularly since it relates to both energy and volume:

$$\varepsilon_h = \left(\frac{\Delta C_p}{\Delta \alpha^*}\right)\left(\frac{v_h}{v_o}\right)(1 - \varphi_1^*). \tag{20}$$

A detailed report on the implications of the data in Table 2b will be published in the near future.

b) Viscosity. As discussed in a previous section on kinetic theories of glasses, the rates of molecular rearrangements in noncrystalline condensed phases take place by a combination of the elementary processes of shear viscosity (η) and diffusion [Eqs. (12) and (13)]. Numerous empirical and theoretical relationships between viscosity, free volume (140, 175, 176) and temperature have been proposed and have been extensively reviewed by Ferry (123) and Bueche (177).

A fundamental quantity relating the basic viscoelastic functions (i.e., storage, loss modulus and compliance, shear viscosity) is the monomeric friction coefficient, ζ_o, which is a measure of the frictional resistance per monomer unit encountered by a moving chain segment. This coefficient, which reflects the magnitude of the relaxation times τ in the transition zone from high to low viscosity is usually calculated from the relaxation spectrum H, as follows (123):

$$\log \zeta_o = 2 \log H + \log \tau + \log(6/kT) + 2 \log(\pi M_o/a\varrho N_o) \qquad (21)$$

where M_o is the repeat unit molecular weight, a is the mean square end to end distance per monomer unit, ϱ is the density and N_o is Avogadro's number. In Table 3 are listed numerical values of $\log \zeta_o$ at the glass transition temperature T_g, and at some specified temperature, 30—100° above T_g [taken from Ref. (123)]. Also included are the temperatures (T_v) at which either the mechanical or dielectric loss tangent reaches a maximum (temp$_{max}$) at a specified frequency of measurement v. It is to be noted that T_v, even when measured at low v, can be considerably higher than T_g by as much as 60°, depending upon a particular polymer. These differences are not surprising when one considers that the phenomena underlying T_v and T_g are distinct. Unfortunately, distinctions between T_v and T_g are seldom been made by the relaxationists and this has resulted in the erroneous assignment of the glass transition temperature for many polymers. Although the shapes of the various viscoelastic functions in the transition zone from high to low viscosity are similar for many polymers (123), their position on the frequency (i.e., reciprocal time) scale varies drastically for different structures. Consequently, ξ_o, whose magnitude reflects τ also varies by several decades at T_g, as shown in Table 3.

A comparison of ξ_o with φ_1^*, the average free volume at T_g derived from Table 2 shows no correlation. For example, polydimethyl siloxane exhibits an exceptionally low ξ_o yet φ_1^* is also on the low side, while Hevea rubber with an ξ_o seven decades higher has a free volume comparable to the silicon rubber. On the other hand, polyvinyl chloride with a monomeric friction factor comparable to Hevea shows a specific free volume twice that of the rubber. Attempts to correlate other thermodynamic parameters from Table 2 with ξ_o have been equally unsuccessful and it appears that whatever molecular features govern the magnitude of the monomeric friction coefficient, they are not reflected in the magnitude of energy and volume changes.

Table 3. Viscoelastic relaxation data for linear high molecular weight polymers

Polymer	Ref.	M_o (g/mole)	a (cm)	T_v (°K)	ν (Hz)	T_g (°K)	$\log \zeta_o$	$T - T_g$	$\log \zeta_o$	φ_1^*
Polyolefins										
Polyethylene	169	28	—	253	100	237	—	—	—	—
				286	1 m	—				
Polypropylene	169	43	—	263	100	259	—	—	—	—
Hevea Rubber	123 (48)	68	6.8	223	1.2 m	200	4.47	100	−6.49	0.014
Polyisobutylene	123 (20)	56	5.9	225	1 m	202	3.47	100	−4.67	0.016
								93	−4.35	
1,4-polybutadiene	123 (14)	54	6.0			172	0.83	100	−6.16	0.021
1,2-polybutadiene	123 (14)	54	7.55			261	2.38	100	−7.01	
								37	−4.11	
Butyl rubber	123 (17)	56	5.9			205	3.57	100	−4.46	0.013
Polyhexene-1	123 (26)	84	7			218	—	100	−5.48	0.023
Polystyrene	123 (21)	104	7.4	389	0.9 m	373	2.06	100	−6.95	0.022
Styrene 23.5%, butadiene 76.5%, copolymer	123 (16)	65.5	6.7			205	—	100	−6.55	
								88	−6.11	
Ethylene 16%, propylene 84%, copolymer	123 (18)	39.9	6.5			242	3.10	100	−6.23	
Ethylene 56%, propylene 44%, copolymer	123 (18)	34.3	5.5		0.7	216	2.40	100	−7.11	
					1					
Polyvinyl halides										
Polyvinylchloride	123 (25)	62.5	6.0	363		347	4.05	100	−7.46	0.026
Polytrifluorochloroethylene	170	116.5		383		352				0.018
Polytetrafluoroethylene	123			397						—
				393						

Table 3 (continued)

Polymer	Ref.	M_o (g/mole)	a (cm)	T_v (°K)	ν (Hz)	T_g (°K)	$\log \zeta_o$	$T - T_g$	$\log \zeta_o$	φ_1^*
Polyvinyl esters										
Polyvinyl acetate	123 (23)	86	6.9	306	2 m	305	4.29	100	−6.23	0.023
	169			336	100 m			(38)	(−2.63)	
Polymethylacrylate	123 (32)	86	6.8	239	1.2 m	276	6.24	100	−6.24	0.017
								(47)	(−3.15)	
Polyethylacrylate	123 (32)	100		268	1.6 m					
Polybutylacrylate	123 (32)	128		241	0.8 m					
Polymethylmethacrylate	123 (11)	100	6.9	393	0.12 m	(378)				0.030
	171			403		335			−4.40	
Polyethylmethacrylate	123 (5)	114	5.9	372	1 m		6.22	100		0.021
Poly *n*-propylmethacrylate	171			343	1 m	300			−4.77	0.020
Poly *n*-butylmethacrylate	123 (6)	142	6.4				3.81	100		0.015
	171			331	1 m	268			−5.18	
Poly *n*-hexylmethacrylate	123 (10)	170	7.5				2.59	100	−0.75	0.016
								30	−5.37	
Poly *n*-octylmethacrylate	123 (8)	198	7.0			253	2.39	100	−2.29	0.015
								45	−4.04	
Polymethoxyethylmethacrylate	123 (44)	144	7.0			293	5.39	100	−4.60	
Polypropenylethylmethacrylate	123 (46)	172	7.0			253	5.50	100	−4.69	
Polydodecylmethacrylate	123 (a)	254		248		208		50		0.017
Polyoxides										
Polymethylene oxide	172	30				(200)				
Polyethylene oxide	123	44				(206)		−122	< −6.4	
Polypropylene oxide	123 (29)	58		213	1 m	198				
Polyacetaldehyde	123 (29)	45		258	1 m	243			< −6.4	

Table 3 (continued)

Polymer	Ref.	M_o (g/mole)	a (cm)	T_v (°K)	ν (Hz)	T_g (°K)	$\log \zeta_o$	$T - T_g$	$\log \zeta_o$	φ_1^*
Polyphenylene oxide	100	92		381	3.5 m	363				
Poly (2,6-diphenyl) phenylene oxide	102	246		503	3.5 m	498				
Miscellaneous polymers										
Polydimethylsiloxane	123	74	6.2			150	−3.60	100	−7.50	0.011
Polycarbonate[a]	123 (37)	254		431	1 m	(416)				0.036
Polyethylene terephthalate	171					422				
Polyacrylonitrile	173	182		371	0.7 m	(337)				
Poly (oxydiphenylene)-	58	(53)		(413)	(4 m)	360				0.02
Poly(pyromellitimide)	174	342		673	110 m	NOT$_g$				
Polyvinylmethyl ether	123 (32)	58		263	2.7 m	(251)				
Polyvinylethyl ether	123 (32)	72		256	1.1 m	(240)				
Polyvinylpropyl ether	123 (32)	86		246	1.1 m	(261)				
Polyvinyl-n-butyl ether	123 (32)	100		241	0.8 m	(217)				
Polyvinyl-i-butyl ether	123 (32)	100		272	1.2 m	(251)				
Polyvinyl-t-butyl ether	123 (32)	100		356	1.7 m	—				

[a] of bisphenol A.

D. Melting

A comprehensive review of crystallization and melting of high polymers has been given by Zachmann (179). Other previous overviews on the subject have been published by Wunderlich (108, 180) and by Ke (38). A list of melting points for 850 polymers[7] has recently been compiled by Miller (181).

In melting studies one differentiates between metastable and equilibrium crystals. Metastable polymer crystals or crystalline regions are generally produced when bulk material is slowly cooled from the melt or by solution crystallization techniques. Under such conditions polymer chains fold successively about every 50—1000 Å depending upon a particular structure, crystallization conditions, temperature of supercooling, etc. Equilibrium crystals of a few polymers have also been grown in which fully extended (or nearly so) chain configurations have been realized (108).

On heating crystalline polymers one observes generally a continuous loss in crystallinity at temperatures far below the melting point. Such a phenomena has been called partial melting (179), and is apparently common to all crystalline polymers. On cooling such a partially molten material it is possible to regain crystallinity by a process which has been termed recrystallization (108) as distinguished from the main and post crystallization. Metastable polymer crystals can also reorganize at temperatures close to their "original" melting point causing shifts in the fusion endotherm. Overall the complexity of crystal morphologies with defects in between as well as in the interior of the crystals, their high degree of metastability, and the wide distribution of crystal sizes result frequently in complicated thermograms and make a quantitative analysis of melting of crystalline polymers very difficult. Some of these problems do not exist with extended chain crystals. On the other hand, in such equilibrium crystals melting proceeds presumably along the chain beginning at chain ends at rates often slower than the conduction of heat of fusion into the crystal. As a result, the polymer crystal interior superheats above the equilibrium melting point; thus, apparently melting takes place at higher temperatures and over a broader range (180).

---

[7] Values reported in this compilation have been collected neglecting details and may be largely in error.

Partial melting can be studied by measuring the decrease in the degree of crystallinity (x) as a function of temperature in such a way as to avoid crystallization effects. The temperature at which $x = 0$ is of interest in that it is close to the "hypothetical" thermodynamic equilibrium melting point (180).

Reorganization, a process by which defect polymer crystals transform either into different crystal morphologies (i.e., orthorhombic polyethylene goes into a hexagonal close packed structure) or smaller crystals with a high surface free energy (i.e., metastable) rearrange into larger ones just prior to fusion can best be observed by heating samples at drastically different heating rates. Here the melting point decreases with increasing heating rates up to a limit where the heating rate surpasses the time scales required for reorganization (zero entropy production path) and the metastable crystals transform directly into the supercooled melt (180).

Crystallization during partial melting can be studied by rapidly quenching samples to a constant, lower temperature and observing the change in x as a function of time. Depending on the time scale of the experiment one can differentiate between "re"crystallization, whereby the original degree of crystallinity is reapproached or "new" crystallization (179), where a higher degree of crystallinity is attained.

Superheating is a phenomena which has so far only been observed in extended chain crystals of polyethylene, polyoxymethylene, polyphenylene oxide (102), teflon, "poly" selenium, polycaprolactam, and recently also in polyethylene terephthalates (183). The phenomena can be observed in DTA experiments by increasing the heating rate. If the crystal can be heated faster than the melt-crystal interface can progress toward the interior, the crystal superheats and melts at a higher temperature.

In the following discussion, the above phenomena will be illustrated with respect to morphology, thermal history, molecular-weight distribution and structure, by means of melting thermograms.

1. Polyethylene

The melting behavior of polyethylene has been investigated to a greater extent that any other polymer and most theoretical and phenomenological considerations of the melting of long chain molecules are based on this material.

a) Bulk Melting (Spherulites). In Fig. 20 is shown the heat capacity of bulk crystallized PE as a function of crystallinity at different tem-

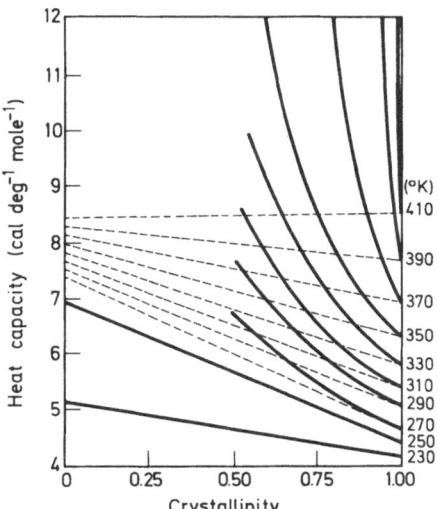

Fig. 20. Specific heat of polyethylene as a function of crystallinity at various temperatures. The dotted line separates the linear region from the nonlinear one. [Ref. (*184*)]

peratures, indicated by solid lines. A linear extrapolation of the C_p curves to zero crystallinity show marked deviations, the lower limits of which are indicated by the dotted curve. Based on the assumption that the specific heat of noncrystalline regions is "essentially" that of the melt (e.g., completely amorphous and crystalline polymer have identical C_p's below T_g), Wunderlich and coworkers (*184*) concluded that the additional contribution of C_p must be due to the presence and interactions of ordered and disordered chain segments in their immediate vicinity. Moreover, it has been argued that above the dotted line, the nature (size and distribution) of disordered regions (defects) increases, thus the observed increase in enthalpy. This growth of defects is apparently reversible at the lower temperatures above T_g and results in additional heat capacities larger than could be accounted for by "vibrations". These added amounts of heat necessary to raise the temperatures have been interpreted by Zachmann as premelting. Several investigators (*185—189*) observed entropy effects and attributed these to added conformations of chains in noncrystalline regions which during melting gain initially more conformational entropy than can be accounted for by the average melt entropy of the material. Consequently, "melting" as observed in Fig. 20 takes place at temperatures far below the equilibrium melting point of 141° C (414 °K) which one observes for corresponding single crystals.

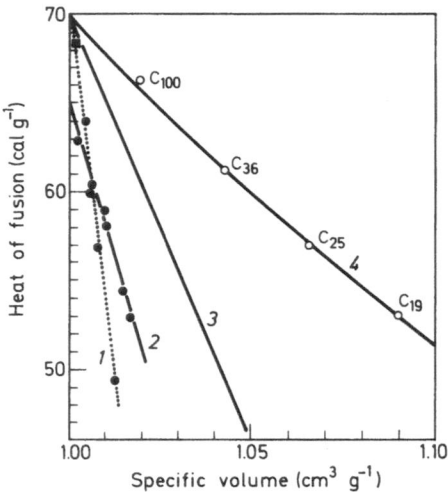

Fig. 21. The effect of crystal morphology on the density and heat of fusion of polyethylene. Curve 1 are polyethylene crystals of the same molecular weight (42,000) grown from paraxylene at various temperatures. From bottom to top: 60°, 70°, and 89.1 °C; Curve 2 represents polyethylene crystals of various molecular weights grown paraxylene at 84.5 °C. From bottom to top molecular weights are: 60,000 (broad distribution) 280,000; 10^7; 140,000; 42,000; and 8,400. Curve 3 shows data on polyethylene crystals reported in the literature by *Fischer*, and *Hendus* and *Illers*. Curve 4 shows data on paraffin crystals of orthorhombic crystal structure. [Ref. (*104*)]

b) Melting of Single Crystals Grown from Solution (Lamellae and Dentrites). Ordinarily, single polymer crystals are grown from dilute solutions and exhibit a more perfect inner structure and a narrower size distribution than melt crystallized polymers. Thermal analysis of these microscopic crystals yields important information about their morphological and thermodynamic nature. Differential thermal analysis has been employed in studies to determine: 1. the melt enthalpy and from it the degree of crystallinity of these so-called single crystals; 2. the effect of crystal thickening on the melt structure (*184*); 3. the relationship of partial melting or reorganization to the observed multiple endotherms in the melting region; 4. the mechanism of new- and re-crystallization following partial melting (*44*); and 5. superheating.

The heat of fusion of 70/cal/g for polyethylene has been determined by extrapolation of the heats of melting as a function of specific volume by Fisher and Hinrichsen (*59*), Hendus and Illers (*190*), Mandelkern and coworkers (*191*), by extrapolation of paraffins to infinite chain length

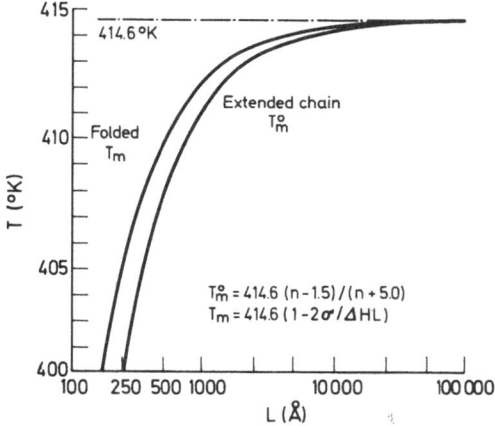

Fig. 22. Melting point of folded chain polyethylene crystals as a function of fold length and of extended chain crystals as a function of chain length. [Ref. (*180*)]

(*192—194*), and from specific heat measurements by Wunderlich (*195*) (see Fig. 21). Melt enthalpies for a variety of polyethylene single crystals gave values between 50 and 60 cal/g (*58*) indicating that the degree of crystallinity for these crystals lies between 70 and 80% depending upon molecular weight, solvent and temperature.

c) **Crystal Thickening.** Isothermal crystal thickening is observed on annealing these crystals above 100°, resulting in an increase in melting point (T_m) according to the simple relationship

$$T_m^\zeta = T_m^\circ \left(1 - 2\,\sigma e/\varDelta h \cdot l\right). \tag{22}$$

Here T_m° is the melting point of an infinitely long, extended chain crystal, σe is the surface free energy and $\varDelta h$ is the specific melt enthalpy. Figure 22 illustrates the relationship between melting point and fold length for high molecular weight material.

d) **Partial Melting and New Crystallization.** Tempered solution grown single crystals exhibit multiple endothermic maxima in the DTA curves (see Fig. 23). The differences in these curves can be explained by considering the following three processes (*59*): Partial or premelting occurs and results in the appearance of the first, smaller peak. Then this fraction of the temporarily molten material crystallizes again resulting in a decrease in the area of the first peak.

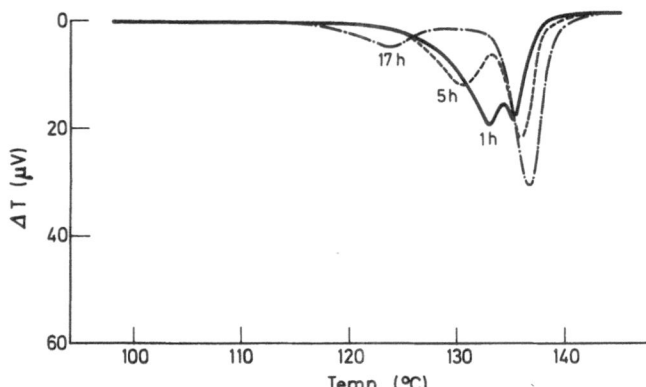

Fig. 23. DTA-Melt curves of polyethylene single crystals, tempered for various time periods at 131.5° C in an air atmosphere. [Ref. (59)]

Finally, the fraction of crystals which remained unmolten increase in thickness (reorganization) causing a shift in the main melt endotherm towards higher temperatures. It is interesting to note that aside from the premelting endotherms which result from quenching, one observes only one main peak. Apparently new crystallization does not proceed through the formation of new crystals but instead the molten fraction crystallizes onto the existing crystals. In fact, X-ray data (59) shows only an increase in the long period indicating that the molten molecules grow parallel to the unmolten chains producing identical fold periods. A shift of the premelt peak towards lower temperatures has been attributed to fractionation (196, 197) during new crystallization. During this process, the undercooling and therefore the rate of crystallization is greater for longer chains; therefore, the remaining molten fraction is enriched with lower molecular weight species which on quenching form lower melting crystals. Alternatively this shift has been attributed to stresses developing in the noncrystalline regions during annealing. Such stressed tie molecules produce a higher entropy in these crystals and therefore lower their melting point.

e) Effect of Heating Rate. The microscopically (hot stage) observable melting point of solution grown, folded chain lamellae decreases smoothly from 130° C at 0.5° C/min heating rate to 120° C for rates above 20° C/min and up to 50° C/min. Zero entropy production is reached at heating rates above 20° C/min (108). Figure 24 shows the effect of heating rate on the DTA main melting peak temperatures for polyethylenes of different morphology.

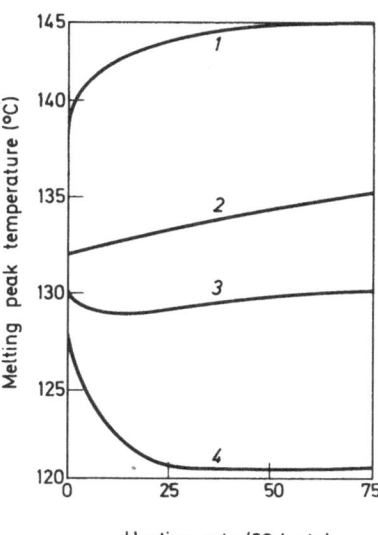

Fig. 24. DTA melting peak temperature as a function of heating rate for polyethylenes of different crystal morphologies. Curve *1*, extended chain crystals;
curve *2*, slowly cooled melt crystallized spherulites; curve *3*, fast cooled melt
crystallized spherulites; curve *4*, solution grown folded chain single crystals.
[Ref. (*108*)]

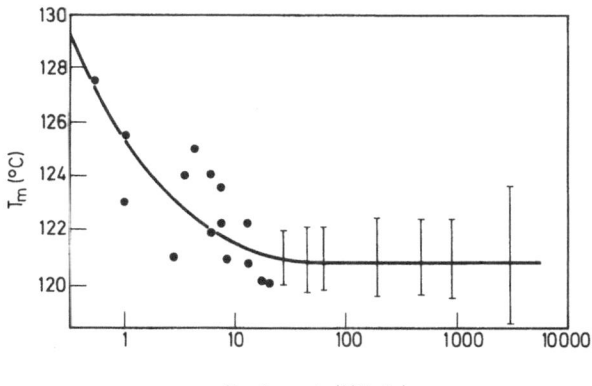

Fig. 24a. Melting point of folded-chain polyethylene single crystals of approxi-
mately 130 Å fold length as a function of heating rate. The bars represent the limits
of a large number of data collected at a particular heating rate. [Ref. (*104*)]

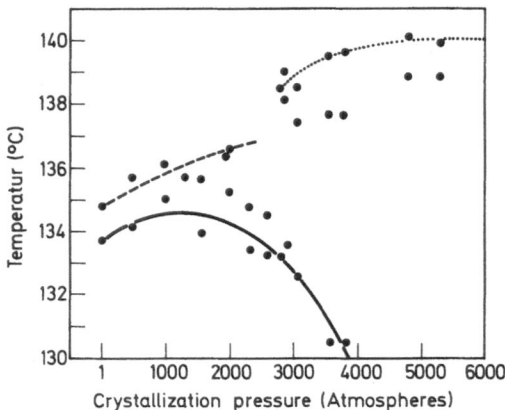

Fig. 25. Experimental maximum melting points of pressure-crystallized polyethylene: (--) folded chain lamellae; (·—·) extended chain lamellae; (○) experimental points. The solid line summarizes the approximate trend of folded chain
lamellae melting peaks. [Ref. (206)]

The melting point of folded chain polyethylene single crystals as
shown in curve 4 has been studied over a wide heating rate range by
Wunderlich (104).

The decrease of T_m with increasing rate which diminishes at fast rates
is typical for reorganization. The limit at about 50° C/min corresponds
to the zero entropy production path for this polymer in which the
metastable folded chain conformations are directly transformed into the
supercooled melt.

f) Effect of Solution Stirring (Fibrillar Structures). Crystallization of
PE by high speed solution stirring produces a fibrillar habit (198) and
consequently preferentially oriented chains parallel to the fiber axis.
These new forms of polyethylene have been subject to thermal analysis
by several investigators (106, 199—201). Annealing these crystals results
in increased crystal perfection and in highly superheatable material (106).
On the basis of superheating, chain orientation and electron microscopy,
it was concluded (106, 199. 200) that these fibrillar structures are
extended chain crystals. Surprisingly however, these crystals melted over
a broad range with DTA maxima of about 137° C (199), a thermal
behavior similar to ordinary bulk crystallized material. Recently Rijke
and Mandelkern (201) have observed a melting temperature of 146° C
for these fibrils, when samples were annealed at 142° C for 19 days (no
comparison was made using bulk crystallized material under those

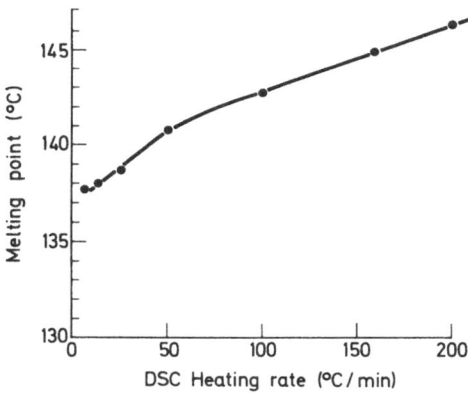

Fig. 26. Effect of heating rate on peak melting point of transparent portion of strands formed at 5 cm/min and 136° C. [Ref. (*209*)]

conditions). These samples apparently exhibit no superheating and lead the authors to believe that high axial orientation in these crystals need not necessarily be identified with extended chain morphologies but instead could result from either an increase in the crystallite size or reduced interfacial free energy.

g) Effect of Pressure (Extended Chain Morphologies). Melt crystallization of PE under pressure yields either thin lamellar crystals (folded chain morphologies) below 2000 atm, extended chain lamellae above 3500 atm, or a mixture of the two forms in the intermediate region (*202—206*). Figure 25 shows the dependence of melting points on pressure for the two types of structures. A slight increase in the melting point of low pressure material was attributed to crystal thickening (*206*). Above 2000 up to about 4000 atm the melting point surprisingly decreased.

This effect was presumably caused by either defects arising from simultaneous growth of extended and folded lamellae or by the presence of thinner folded chain crystals in this region (e.g., surface free energy contributions). The high pressure material exhibit nearly equilibrium melting behavior and showed that almost perfect extended chain crystals were present.

The combined effects of pressure and orientation on the melting behavior of PE have recently been investigated using a pressure capillary viscometer (*207*). Although pressures of only 1900 atm could be realized in this study, the melting points and superheatibility (see Fig. 26) of this

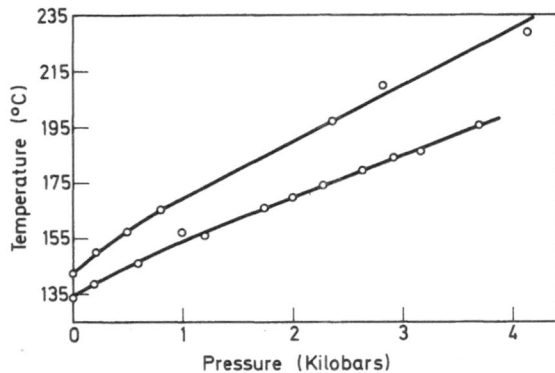

Fig. 27. Melting curves of extended-chain and folded chain polyethylene: (upper) extended-chain polyethylene; (lower) folded-chain polyethylene [Ref. (208)]

oriented material were attributed to the presence of extended chain crystal components. Differential thermal analysis of folded and extended chain crystals under pressure up to 4200 atm also was performed to determine the effects of crystallization and melting (208).

From the data in Fig. 27 it is seen that the higher melting temperatures of extended chains persist at elevated pressures and that these crystals remain stable at temperatures above the melting range of folded chain configurations.

h) Other Polyolefins. With the advent of differential scanning calorimetry has come reinvestigations of the melting behavior of many previously studied homo- and copolymers, and it is beyond the scope of this review to consider them all in detail.

In a recent reinvestigation on multiple melting in natural rubber (209), a study of the effect of the crystallization temperature on the melt behavior of this material was undertaken. An endothermic maximum at 260 °K, about 20° below the main fusion endotherm has been attributed to the melting of thermodynamically less stable crystallites (higher interfacial free energies) which apparently develop during isothermal crystallinization at higher levels of crystallinity. Melting of isotactic polystyrene has been reinvestigated by DSC and solubility measurements (210).

Depending upon the melt crystallization temperature 1, 2 or 3 melt endotherms occur (Fig. 28). At large supercooling, crystallization from the melt produces a small melt endotherm just above T_c, presumably due to secondary crystallization of melt trapped within the spherulites

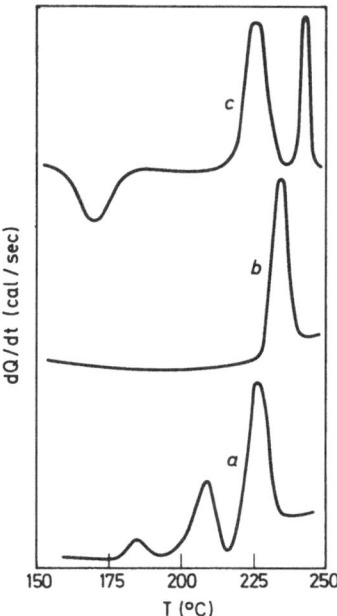

Fig. 28. DSC melting curves of isotactic polystyrene crystallized in the melt at different crystallization temperatures T_c: a 170° C; b 215° C; c 233° C. Heating rate 8° C/min. [Ref. (210)]

(curve a). At 210° C < T_c < 225° C, a single melt endotherm related to the primary crystallization process is observed (curve b), from which an equilibrium melt temperature of 242° C was found by extrapolation of $T_m - T_C$ curves. The origin of double melting (i.e., peak 2 and 3) (curve a and c) was found to result from recrystallization, an observation which was also made in an earlier study by Pelzbauer and Manley (118).

Other recent detailed studies of melting have been published on trans-polychlorophene (211), on trans-1.4-polyisoprene (212) and on chlorinated polyethylenes (119).

2. Polyalkyl Ethers

By far the most extensively studied homolog in the polyalkyl ether family is the commercially important polyoxymethylene (POM) and a wealth of melting data can be found in the literature on this material

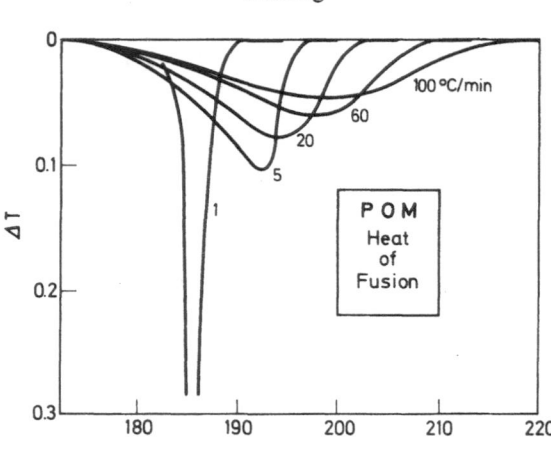

Fig. 29. Normalized DTA-traces at different heating rates for extended chain POM fibers. ΔT is plotted versus the reference temperature T_R. The rectangle represents the normalized heat of fusion A. [Ref. (182)]

[see Ref. (182)]. The variety of crystal types possible make this polymer particularly amendable to melting studies. Extended chain crystals (fibers) with diameters of several hundred microns and lengths of over a centimeter can be obtained by simultaneous polymerization and crystallization of either trioxane or formaldehyde. Single crystals in the form of thin (60—100 Å) chain folded hexagonal lamellae can be grown from a variety of solvents over a wide temperature range. Spherulitic morphologies are readily obtained on supercooling melts with dimensions depending upon T_C. Hedritic geometries have been grown from the melt and dentrites can also be formed from dilute solutions at large supercooling. A DTA trace of extended chain POM fibers is shown in Fig. 29. Clearly superheating of about 20° C in these crystals is evidenced by the increase in melting maximum with increasing heating rate. The zero entropy production melting point from this data lies at 182.5° C and is likely to be close to T_m°, the thermodynamic equilibrium melting point. The fact that superheating or reorganization has often not been recognized in polymer melt studies has led to misinterpretation of thermograms and false assignments of melting points in the past.

A DTA curve of solution grown folded chain crystals is shown in Fig. 30. The three processes — premelting (1st sharp endotherm), recrystallization (DTA overshoot) and melting (2nd endotherm) are clearly

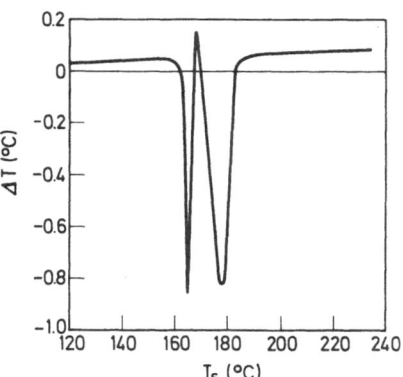

Fig. 30. DTA trace (ΔT versus T_s) of solution grown POM hedrites measured at 10° C/min heating rate. [Ref. (*182*)]

distinguished in this curve. The first peak in the hedrites lies about 5° lower than for single crystals and presumably is due to a smaller fold length (*182*). Other polyalkyl ethers whose melt behavior has been investigated in detail are polyethylene oxide (*213—216*) and polypropylene oxide (*217—220*).

Employing a dilatometric technique, Booth and coworkers determined the melt behavior of fractionated PPO as a function of crystallization temperature and by this technique found the thermodynamic melting point for PPO to be near 82° (Fig. 31), surprisingly 100° C lower than that of POM. These authors also detected three dilatometric transitions for partially isotactic material. The first below 60° was ascribed to the melting of lamellar crystals of limited thickness arising from the isotactic sequence. The other two transitions which depended on T_c were attributed to the melting of lamellae with thicknesses determined by primary nucleation.

3. Polyamides

Melting thermograms of numerous polyamides have been determined by Ke and Sisko (*221*) in a qualitative fashion to determine the effects of polymer composition on melting points (Fig. 32).

The thermograms are characteristic of semicrystalline materials. In all cases small baseline deflection at about 50° C due to glass transitions

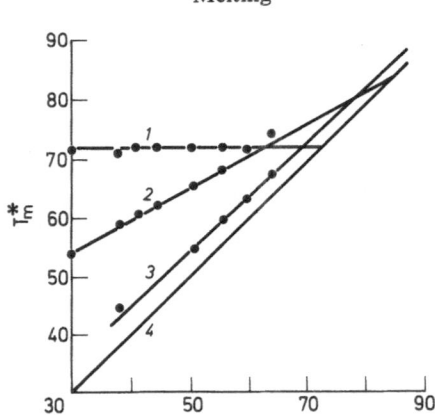

Fig. 31. Melting points of poly(propylene oxide) fraction (T_m^*, °C) vs. crystallization temperature (T_c, °C): $1\ T_1^*$; $2\ T_2^*$; $3\ T_3^*$, $4\ T_m^* = T_c$. [Ref. (217)]

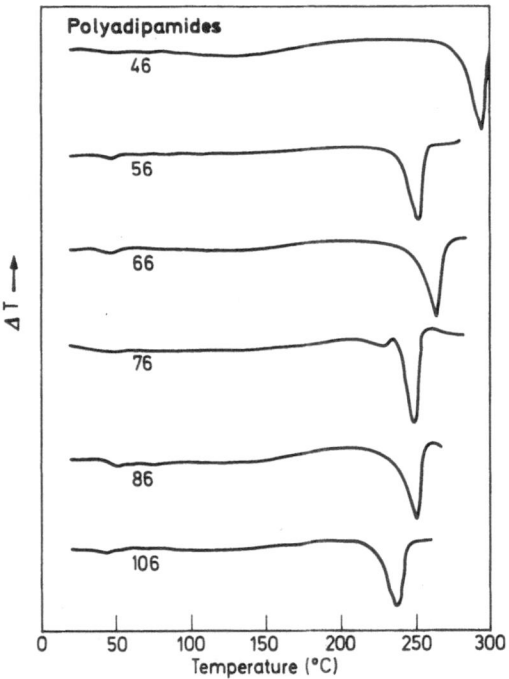

Fig. 32. Melt thermograms of polyadipamides [Ref. (221)]

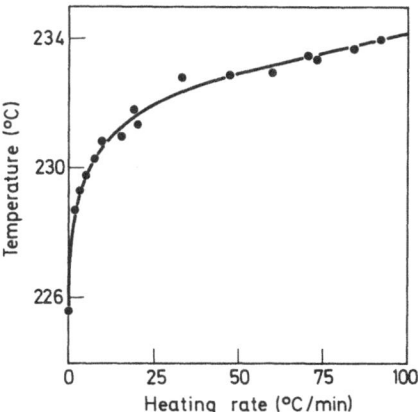

Fig. 33a. DTA melting peak temperatures of zone-polymerized ε-caprolactam as a function of heating rate. The zero rate heating point was obtained by scanning calorimetry. [Ref. (222)]

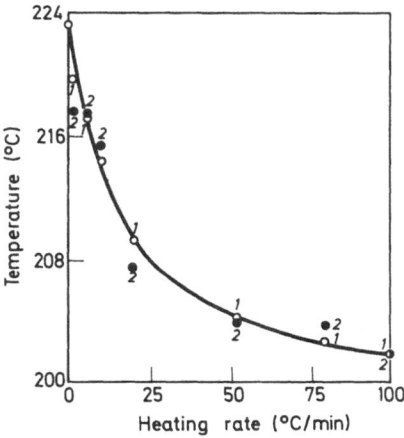

Fig. 33b. DTA melting peak temperatures of solution crystallized polycaprolactam as a function of heating rate: 1 initial melting points; 2 remelting after cooling at 8° C/min. The zero-heating-rate point of the initial melting was obtained by scanning calorimetry. [Ref. (222)]

in the amorphous portion of the polymers are apparent. The effects of repeat unit structure on the melting points within a homologous series were also demonstrated, showing a zigzag relationship between melting

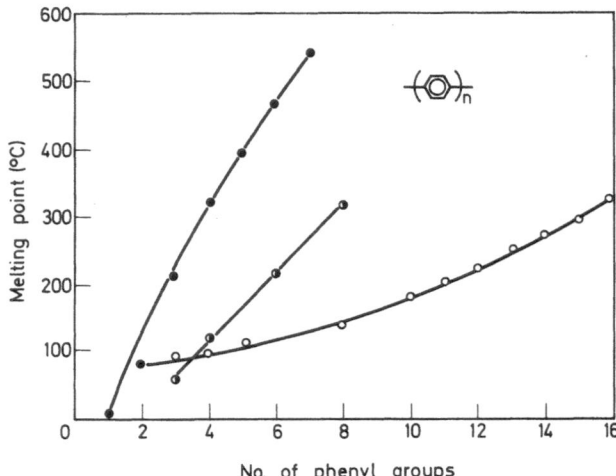

Fig. 34. Melting points of aromatic polyphenyls (● para, ○ meta, ◑ ortho)

point and number of carbon atoms in the chains. In all cases the even numbered diamines produce polymer structures with high melting points. A detailed analysis of melting in polycaprolactam was recently performed by Liberti and Wunderlich (222). Four different morphological structures were produced by different techniques [zone polymerized ($T_m = 225.6$), solution crystallized ($T_m = 200°$ C), annealed ($T_m = 222°$ C), and melt crystallized ($T_m = 255°$ C)], all showing different melting behavior. Zone polymerized caprolactam exhibits superheating and has a "zero-heating-rate" melting point of 225.6° C (Fig. 33a), close to melt annealed commercial PCL (m.p. 223.6°). However, the heating rate dependence on melting of the latter is similar to that of the solution crystallized polymer (Fig. 33b), showing premelting (or reorganization) at lower heating rates and zero entropy melting at about 50° C/min heating rate.

4. Aromatic Polymers

Among the very large number of chemically different aromatic and aromatic-heterocyclic polymers synthesized within the last decade, only a few exhibit any crystallinity and therefore have melting points. Among those which do, are the polyphenylenes, polyphenylene ethers, poly-carbonates and aromatic polyamides.

a) **Polyphenylenes.** Of the polyphenylenes, only lower molecular weight species exhibit melting prior to thermal decomposition. In Fig. 34 are plotted the melting points of the three possible isomeric oligomers as a function of the number of phenylene moieties in the chains. Melting in paraphenylene homologs appears limited to $n = 7$, probably because the higher homologs decompose prior to melting. Both the para and ortho compounds melt higher for a given n than the meta isomer. There are several reasons for these differences in melting behavior. First, both para and ortho isomers possess in addition to their regular resonance structures also a form in which resonance-overlap between ring structures is possible:

These extra resonance forms will contribute to the average structure in the ortho and para isomers and in fact in the crystal may be the dominating electronic configurations. The result is 1. a higher stiffness energy in structures a and b, and 2. coplanar chains, resulting in a better crystal packing geometry. Structure b should exhibit the highest crystal perfection and give the highest melting points. Secondly, in the meta

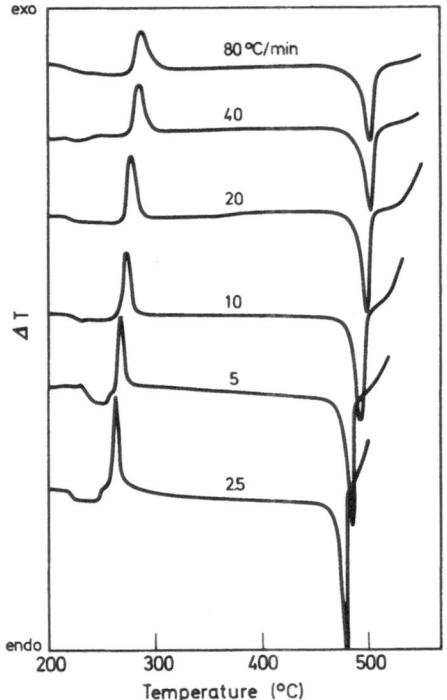

Fig. 35. Effect of heating rate on cold crystallization, melt transition, and thermal decomposition temperatures. [Ref. (*102*)]

isomer resonance overlap is not possible; a conformation such as *d* (although otherwise identical to *c*) could be frozen into the crystal structure and give rise to less perfect and therefore lower melting crystals. It is of interest to note the upward curvature in the meta oligomers. If such a trend would continue as *n* increases it would lead to infusable polymers at a relatively low degree of polymerization.

b) Polyphenylene Ethers. Melting of polyphenylene ethers has been studied in different ways by several investigators (*100, 102, 155, 156, 225—229*). Repeat unit structure and some thermal properties of three representative polymers are given in Table 4. The unsubstituted poly (1.4-phenylene ether) can be crystallized under a variety of conditions (see Table 1) to a high degree. On the other hand, bulk crystallization of the 2.6 diphenyl homolog can only be achieved in the solid state below T_m as the polymer decomposes on prolonged heating in the melt. The commercially important 2.6-dimethyl compound is very difficult to crystallize to a high degree of perfection, which is probably the reason for its low T_m and high T_g/T_m ratio. It has been suggested that steric hindrance caused by methyl groups produces a relatively loosely packed crystal (229) and therefore also a low heat of fusion. Rate dependent melting of structure 3 (Table 4) over a range from 2—80° C was performed by DSC (Fig. 35). It was found that virgin polymer superheats by 5° (Fig. 36, after correcting for thermal lag), suggesting either the presence of extended chain crystals, or metastable morphologies in which the amorphous conformations are temporarily restricted (184, 186), thus causing an initially smaller entropy gain on melting.

Table 4. Effect of structure on the thermodynamic properties of poly(phenylene ethers)

Structure	T_g(°K)	$(\Delta C_p)T_g$, (cal/deg per segment)	T_m(°K)	X_c(%)[a]	T_g/T_m	$\Sigma(\Delta S_f)$, (cal/deg mole)	ΔH_f (cal/mole)
—⟨O⟩—O—	363	2.76	535	72	0.69	3.50	1870
—⟨O⟩(CH₃)(CH₃)—O—	480	3.45	535	40	0.90	2.18	1290 (229)
—⟨O⟩(C₆H₅)(C₆H₅)—O—	493	4.11	757	52	0.69	3.85	2916

[a] X-ray crystallinity [Ref. (100)]

A zero entropy production path is indicated at a heating rate of 40°/min by a drastic change in slope of the melting point vs. heating rate plot. The melt endotherms, even at very fast heating rates, are exceptionally sharp, suggesting a narrow crystallite size distribution. Electron microscopy of this "cold crystallized" material showed the presence of single crystals (102), which to the writer's knowledge have previously not been observed in any other bulk crystallized polymer.

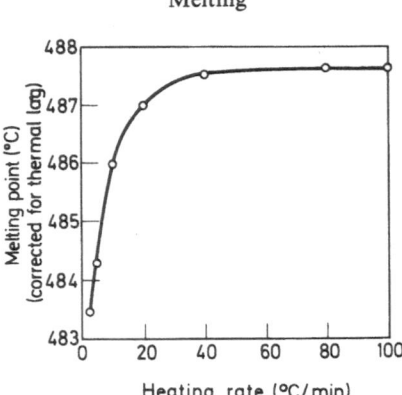

Fig. 36. Superheating in poly (2,6-diphenyl) phenylene ether

Exceptional about the behavior of this polymer is also the insensitivity in melting point with crystallization conditions. Samples crystallized during heating at rates between 2 and 80°/min exhibit surprisingly similar melt thermograms (Fig. 35).

5. Polyesters

Numerous studies on polyester melting have been published (*230* to *236*). More recent reports can be found in papers by Hobbs and Bill-meyer (*237*), Miyagi and Wunderlich (*183*), and Sweet and Bell (*238*), Roberts (*239*), Ikeda (*240*) and Holdsworth and Turner-Jones (*241*).

Superheating was observed in polyethylene terephthalate (*183*) and attributed to entropic restrictions of the partially molten crystals (Fig. 37). Since no extended chain conformations were found to be present, a mechanism was proposed by which partially extended "tie" molecules temporarily restrain the metastable crystallite network from collapsing at the higher heating rates.

PET exhibits multiple endotherms in the melt region similar to polystyrene and other polymers (Fig. 38), depending upon its thermal history.

A decrease in the premelt endotherm and a lowering in its peak temperature as the cooling rate is decreased is indicative of reorganization (i.e., premelting followed by recrystallization).

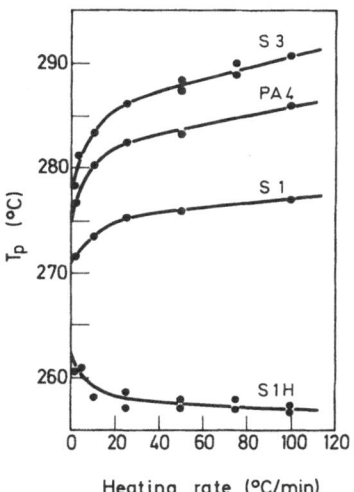

Fig. 37. Melting peak temperatures of PET as a function of heating rate by DTA: *S1* sample crystallized at 250° C for 48 h; *S3 S1* annealed at 260° C for 48 h; *S1H S1* hydrolyzed at 210° C for 1 h (oligomer sample); *PA4 S1H* annealed at 250° C for 24 h. [Ref. (*184*)]

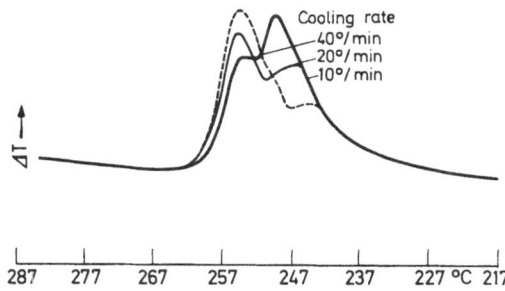

Fig. 38. Cooling rate effect on observed PET endotherm areas and peak temperatures. The sample was remelted at 10° C/min after each different cooling rate. [Ref. (*238*)]

Oligomeric polyesters containing p-phenylene moieties have recently been studied to ascertain the role of aromatic rings in the formation of liquid crystals (*242*).

Table 5. Transition temperatures for p-phenylene p-alkoxybenzoates (I)

$$RO-\langle O\rangle-CO_2-\langle O\rangle-CO_2-\langle O\rangle-OR$$

Compd	Transition temp (°C)
Ia R = CH$_3$	solid $\xrightarrow{213}$ nematic $\xrightarrow{297}$ isotropic
Ib = C$_2$H$_5$	solid $\xrightarrow{226}$ nematic $\xrightarrow{287}$ isotropic
Ic = n-C$_3$H$_7$	solid $\xrightarrow{175}$ nematic $\xrightarrow{249}$ isotropic
Id = n-C$_4$H$_9$	solid $\xrightarrow{153}$ nematic $\xrightarrow{241}$ isotropic
Ie = n-C$_6$H$_{13}$	solid $\xrightarrow{121}$ nematic $\xrightarrow{211}$ isotropic
If = n-C$_u$H$_{15}$	solid $\xrightarrow{121}$ nematic $\xrightarrow{198}$ isotropic
Ig = n-C$_8$H$_{17}$	solid $\xrightarrow{118}$ nematic $\xrightarrow{192}$ isotropic

[Ref.(242)]

It was demonstrated that aromatic rings are essential in the formation of neumatic crystals whose transition temperature could be varied over a 100° range depending upon the alkyl residue in the chains. The formation and stabilization of mesophases in these oligoesters has been attributed not only to the rigid linear geometry of aromatic rings, but also to intermolecular effects arising from the polarizability of π-electrons leading to increased intermolecular dipole-dipole attractions.

It should only be a matter of time before mesophase type anisotropies will also be discovered in polyesters containing phenylene rings.

6. Melting of Copolymers

The incorporation of a comonomer into a polymer structure generally lowers the melting point to an extent which depends among other things on the morphology of the copolymer. For comonomers of similar structure it is possible that both fit into the same crystal lattice in which case the crystal may only be slightly disrupted by the copolymerization (177). If the comonomers are dissimilar, the crystal may entropically segregate all foreign monomer units and thereby strongly affecting the melt behavior of the copolymerized structure.

An important type of copolymer is one in which the chain structure consists predominantly of one particular unit (A), such as in methyl ethyl, acetyl, halogen or branched polyethylenes. An approximate theory for the melting of this type of copolymer has been worked out by Flory (244), according to which

$$1/T_m = 1/T_{m0} - (R/\Delta H_f) \ln X_A \tag{23}$$

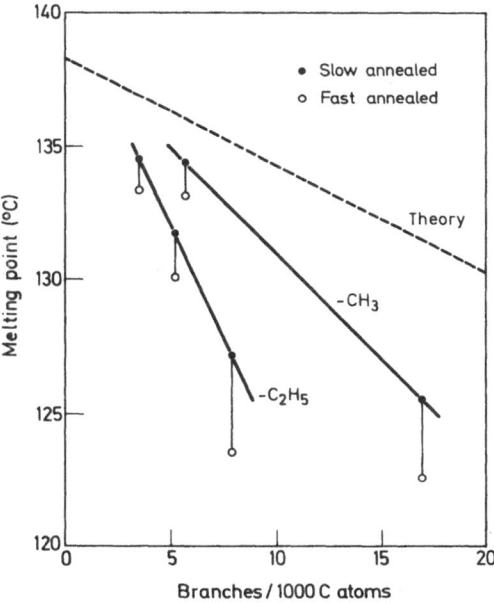

Fig. 39. Melting points of methyl- and ethyl-branched polymethylenes. [Ref. (38)]

where X_A is the mole fraction of the parent polymer, whose melting point is T_{m0}.

The effect of a random incorporation of methyl and ethyl pendant groups into polymethylene chains is shown in Fig. 39. The melting points predicted from Eq. (21) are indicated by the dashed line, and those experimentally determined are represented by the solid lines. The discrepancy between theory and experiment may: 1. in part be a reflection of the inadequacy of the theory; 2. possible also be due to the inability to detect accurately the DTA-melting peak for these broad melting copolymers (245); 3. be caused by discrepancies in the heats of fusion (ΔH_f) computed by different techniques; and 4. be due to higher surface free energies at higher branch concentrations.

An example of copolymer melting in which both components are present in substantial fractions is illustrated in Fig. 40 for the "interfacial" adipamide-sebacamide copolymer system. When the DTA-peak temperatures were plotted against copolymer composition, a phase diagram is obtained in which the eutectic melting point of 79° corresponds to a mole fraction of 70% sebacamide. With the exception of the eutectic, the copolymer melt thermograms showed multiple peaks, which

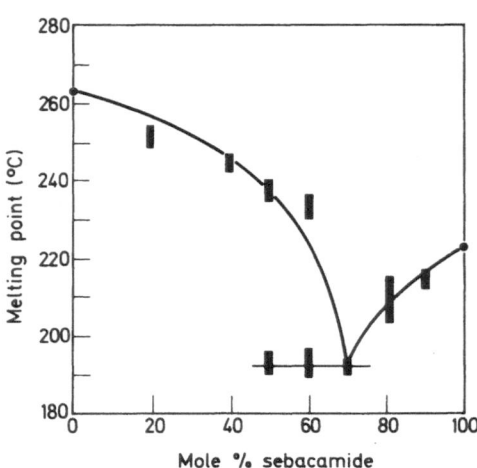

Fig. 40. Melting points of adipamide-sebacamide copolymers. [Ref. *(221)*]

were ascribed to crystalline regions of segregated adipamide and sebacamide blocks. That not all copolymers exhibit eutectic behavior was
demonstrated with the system adipamide-terephthalamide *(221)*. Here
the melting points rise gradually with increasing terephthalamide
content (i.e., chain stiffness), a phenomena which was attributed to "iso-
morphism" of the two blocks.

Another illustration of a copolymer type effect is shown in Fig. 41,
where the melting points of seven representative polymer families,
encompassing 50 different structures, have been plotted as a function of
the number of consecutive carbon atoms in the chains. Here the chains
are presented by "ordered" methylene "co" segments increasing in
sequence and periodically spaced between heteroatoms (i.e., sulfide,
oxide) or groups of atoms (amide, ester, etc.). The melting points of the
polymers decrease as the methylene "comonomer" content increases and
by extrapolation merge at 137° C at which point the number of consecutive methylene groups reached about 20 regardless of the nature of
the heteroatom. It is probably more than a mere coincidence that this
temperature corresponds to the experimental melting point for carefully
crystallized linear polyethylene. What is suggested in the data of Fig. 41 is
that intermolecular attractions such as hydrogen bonding (in polyamides) and dipole-dipole associations (in polyesters) which apparently
dominate the melting behavior at low n vanish and cease to contribute
as n approaches 20.

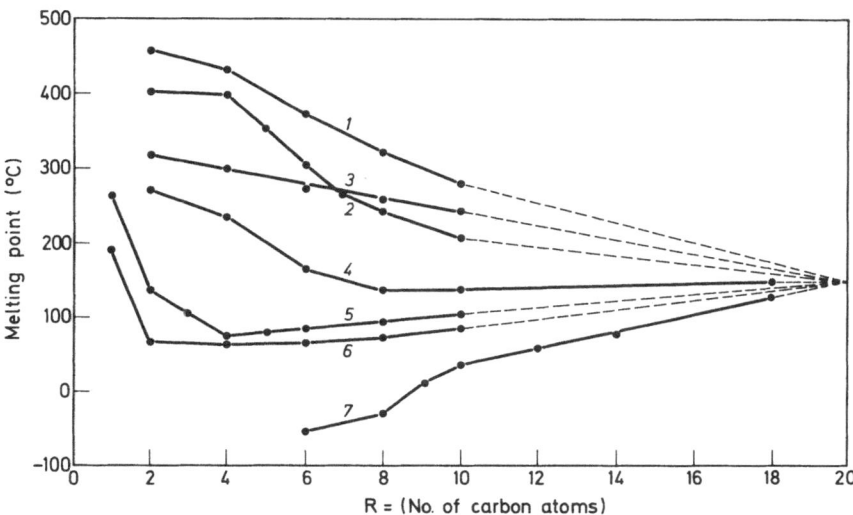

Fig. 41. Melting points of polymers as a function of methylene groups in the chain *1* polyterephthalamides, *2* polycarboamids, *3* polyadipamides, *4* polyterephthalates. *6* polyethers, *7* poly(1-alkenes)

7. Polymer-Blends

Mixing of polymers either of melts or from solution are old techniques for obtaining specific mechanical or physical properties. The intermediate thermomechanical properties obtained are frequently desirable for a particular process or end use. Aside from numerous qualitative studies, very little pertinent work has been reported on the thermophysical behavior of polyblends.

DTA curves of commercial high and low molecular weight polyethylene, polyethylene-copolypropylene and polyamide blends as well as mixtures of polymers and low molecular weight compounds have been discussed by Ke (*38*).

E. Crystallization

Long range three dimensional order in polymers occurs during parallel alignment of either whole molecules as in extended chain crystals or relatively long chain segments as in periodically folded chain structures. The single most important factor dictating the crystallizability of polymers is conformational regularity, the ability of chains because of their molecular constituents to conform to a regularly repeating sequence (i.e., stable helix, planar zigzag, etc.). Without such molecular geometric constraints a polymer cannot crystallize. In systems capable of stereo-isomerism, additional configurational requirements (tacticity) must be satisfied to attain conformational order. Because of the statistical nature of macromolecules, it is douptful that perfect conformations will ever exist and consequently completely developed, perfect polymer crystals may experimentally never be achieved.

The extent to which crystallization in high molecular weight componds occurs depends on both kinetic and thermodynamic factors. Based on equilibrium considerations, Flory (243), utilizing the lattice model, predicted several decades ago that high polymers should attain a high degree of crystallinity. That this is generally not the case can simply be attributed to: 1. various kinds of lattice defects, and 2. rate effects. Numerous recent statistical and phenomenological thermo-dynamic (245, 247, 250) treatments and kinetic theories (244, 246, 248) have been developed in which different view points (and controversies — 249) of polymer crystallization have been examined.

The much debated nucleation theories were originally developed to explain thermal analysis data such as the strongly negative temperature coefficients of growth rates, and the dependence of chain folded lamellar thicknesses on the crystallization temperature (244). In the following discussion, some recent observations on growth velocities, isothermal thickening, undercooling and nonisothermal aspects of crystallization will be emphasized.

1. Degree of Crystallinity

In addition to X-ray, density and infrared techniques, DTA is becoming increasingly popular in determining the degree of crystallinity of polymer samples (44, 251—255). In analogy to the concept of additivity of specific volumina of amorphous and crystalline phases (i.e., $V = \alpha V_c + [1 - \alpha] V_a$), the calorimetric determination of crystallinity is based on the presumption of additivity of enthalpies:

$$H = \alpha H_c + (1 - \alpha) H_a \tag{24}$$

where α is the mass-fractional crystallinity, H_c and H_a are the enthalpies of the crystalline and amorphous phase respectively. The degree of crystallinity (α) at temperatures below the onset of melting is obtained from the following equation:

$$\alpha = \frac{\Delta H}{\Delta H_c} \tag{25}$$

where $\Delta H = H_a - H$ (= melt enthalpy) and $\Delta H_c = H_a - H_c$ (= heat of fusion). For small crystals (i.e., most polymers), the measured heat of melting is reduced by the surface free energy:

$$\Delta H_{exp} = \alpha(\Delta H - 2\,\sigma_e v_v/\xi) = \alpha\,\Delta H \xi \qquad [\text{Ref. (44)}]$$

where σ_e is the surface enthalpy of crystals of thickness ξ with a specific volume v. Once the relationship between the experimentally accessible melt enthalpy ΔH_{exp} and specific volume v are established, the other important quantities α, ΔH, σ_e and ξ in the above equation can in principle be computed from the following equations:

$$\xi = L \alpha v_c/v_a \tag{26}$$

where L is the long period determined from X-ray data.

$$\alpha = \frac{\Delta H_{exp} + \dfrac{2\sigma_e v}{L}}{\Delta H} \tag{27}$$

For polyethylene, a linear relationship between melt enthalpy and specific volume is obtained as shown in Fig. 42, regardless of what crystallization conditions were employed.

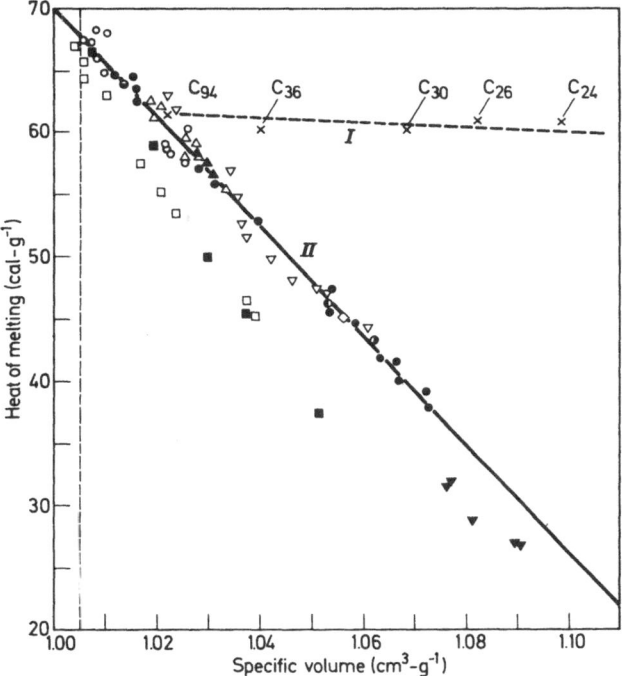

Fig. 42. Melt enthalpy of polyethylene as a function of specific volume. [Ref. (252)]. (○ PE isothermally crystallized at pressures of 1000 to 4000 atm, ● slowly crystallized from the melt, ▼ solution crystallized)

From the slope and intersect of the line obtained in Fig. 42, Hendus and Illers derived the following equation for polyethylene

$$\Delta H = \Delta H_c - 437 \, (v_{20} - 1) \tag{28}$$
$$= -333 \, (v_{20} - 1).$$

While the computation of α from Eq. (25) is in very good agreement with the density method (44, 252), calculations of σ_e, the surface free energies for polyethylene samples has resulted in large statistical errors presumably because the contribution of σ_e to ΔH_{exp} is relatively small even for microscopic crystals. Thus, a small deviation in ΔH and v produces a considerable error in σ_e (44). A better estimate of σ_e is obtained from Eq. (22).

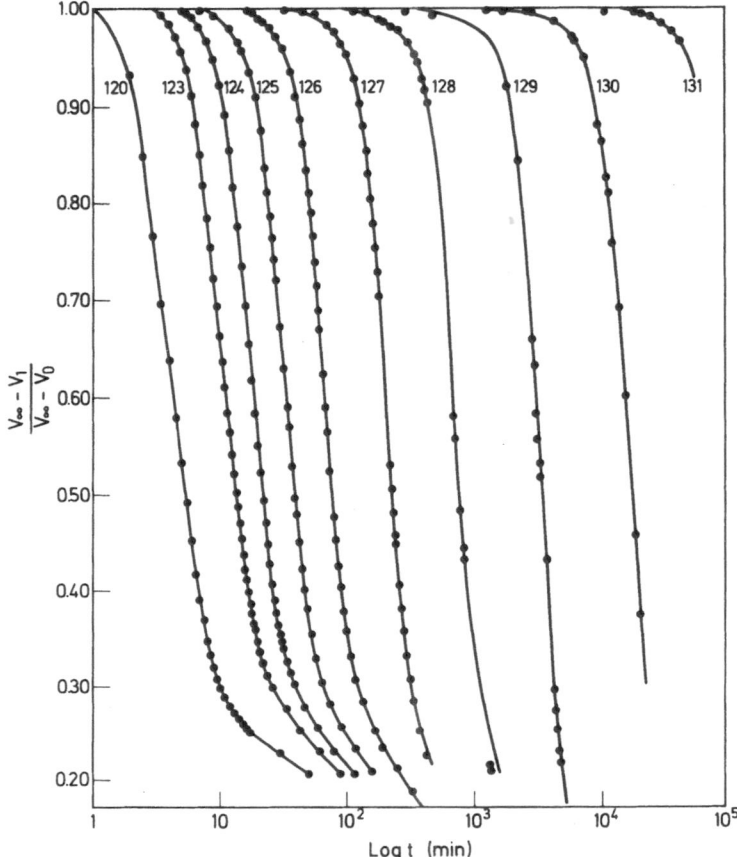

Fig. 43. Plot of quantity $(V_\infty - V_t)/(V_\infty - V_0)$ against the logarithm of time for the crystallization of a linear poly(ethylene). Temperature of crystallization is indicated for each isotherm. [Ref. (*246*)]

2. Growth Velocities

For reproducibility of crystallization rates it is important that 1. polymer samples are purified to avoid heterogeneous nucleation; 2. degradative processes are eliminated by proper choice of protective atmospheres, etc., and 3. one starts from the true liquid state (*246*).

Isothermal bulk crystallization is generally followed either dilatometrically or microscopically and a wealth of data has been reported in this way (*116, 246, 256*). For a first approximation one applies the Avrami equation:

$$\chi = 1 - \exp(-Kt^n) \tag{29}$$

where $n = 3$—4, depending upon the nature of nucleation. Reasonable agreement between Eq.(29) and experimental results can often be achieved over longer periods of time by adjusting n (*257*).

Figure 43 is a typical example of a dilatometrically determined crystallization rate for polyethylene, from which all the principle features of the "primary" crystallization process (spherulitic growth to its termination) can be deduced (*246*).

Thermal analysis curves such as shown in Fig. 43 by no means tell a complete story. Important secondary [for a definition of "secondary", see Ref. (*258*)] rate processes occur in polymers which may continue over many decades of logarithmic time (*258, 259*), and exert a profound influence on the detailed morphological structure and therefore on the mechanical properties of polymers. Unfortunately these secondary rate processes are inaccessible to observation by conventional techniques (*258*) and so far their description has probably been more fantasy than fact. At present, secondary (or post) crystallization is envisioned to involve the ordering of originally segregated smaller molecular weight fractions (*258*) of noncrystalline domains into various morphologies (including chain folded and extended chain crystals) and are accompanied simultaneously by internal reorganization of those structures formed during primary crystallization (*260, 261*).

3. Isothermal Thickening

A consequence of post or secondary crystallization is a thickening in the long period of polymer crystals. As already noted in Section D.1c this effect can be studied over relatively short periods of time (i.e., several hundred hours) by annealing semicrystalline material. In Fig. 44 this effect is demonstrated for solution grown polyethylene single crystals.

On annealing these partially crystalline single crystals for short periods, L increases at first (t_o) rapidly up to about 250 Å (L_o) and from there on varies with the logarithm of time according to the equation:

$$L = L_o + B(T) \log (t/t_o + 1) \tag{30}$$

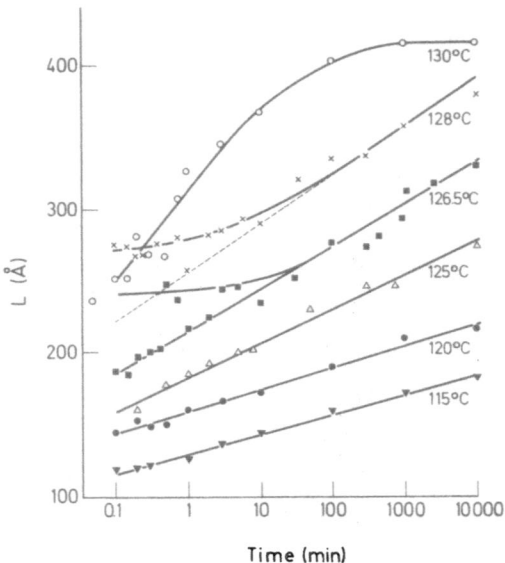

Fig. 44. X-ray long period as a function of time with annealing temperature as a parameter. [Ref. (262)]

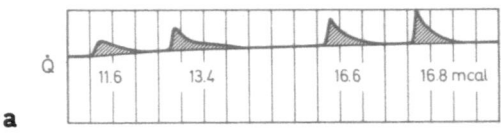

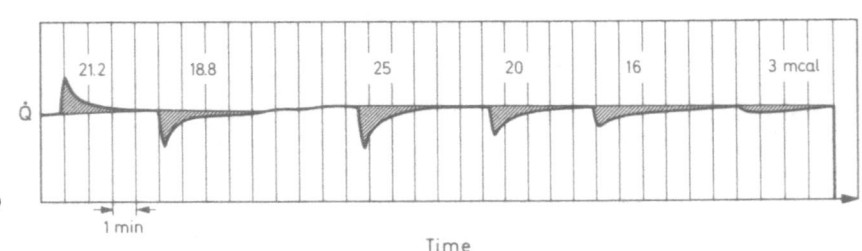

Fig. 45. Heat effects during successive elongation a) and retraction b) of vulcanized rubber. [Ref. (262)]

where the proportionally constant B depends strongly on temperature (262). Even at higher temperatures (i.e., 130° C) during melting one notes thickening of the newly formed crystals, reaching an apparent limiting value at 400 Å. Similar behavior is observed for melt crystallized material (256, 266). Based on these data and supporting evidence from DTA results (see Fig. 23), Fisher and coworkers have concluded that the mechanism of crystal thickening proceeds not simply via a two-phase new crystallization (263, 264) following partial melting, but in a special mode which involves irreversible structure changes in defect regions. While this general physical view is experimentally and theoretically well substantiated, elucidating the fine morphological details for these apparently irreversible thermal processes will not be a simple problem.

Crystal thickening has also been observed on polyoxymethylene (267), in polyethylene oxide (269, 270), and polyesters (269).

4. Orientation Effects

Ordinary thermal processing of plastics almost invariably produces some orientation effects. Depending on the end use of a material such effects can be advantages as in melt spinning of fibers or extremely undesirable whereever isotropic performance is required. Many studies on the effects of orientation on the properties of polymers have been reported and reviewed (38, 92, 271—273).

Well known examples of orientation induced crystallization are the stretching of rubber, fiber orientation in polyamides and polyesters, and the biaxial stretching of polycarbonate films. An example of thermal effects during induced crystallization in rubber is shown in Fig. 45, where the heat flow (Q) for successive steps of elongation and retraction is plotted as a function of time.

Since the amount of mechanical work for each step was identical, the exotherms during stretching increased steadily with the degree of elongation (crystallization) and the endotherms decreased steadily on retraction (remelting).

DTA has also been employed in orientation studies of polyethylene terephthalate (271), polyethylene fibers (273) and polypropylene (274). A dramatic illustration of how orientation in amorphous polymers can affect the resulting DTA curves is shown in Fig. 46.

At first sight it would appear that the stretched material (curve a) exhibits a smaller heat of crystallization than the isotropic sample (curve b). However, it was shown by the Bayer workers on samples which were

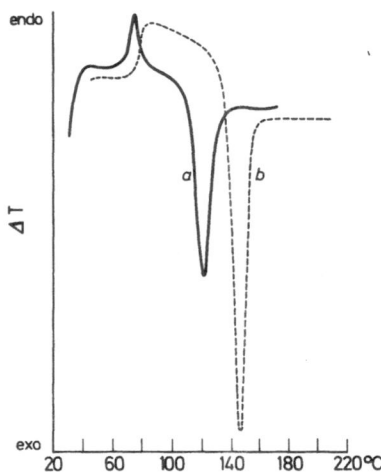

Fig. 46. DTA curves for oriented-amorphous a) and isotropic-amorphous b) PET film. [Ref. (271)]

restrained during the measurement that this apparent decrease in the heat of crystallization is merely masked by an overlapping endothermic heat of shrinking. The drastic lowering of the crystallization peak temperature from 145—120° C has been attributed to local ordering which occurs during deformation presumably by chain slippage, resulting in a three dimensional "paracrystalline type" chain assembly. Another demonstration of the usefulness of thermal analysis in morphology studies is given in Fig. 47. Biaxial stretching of polyethylene film produces two populations of crystals differing in orientation, which as shown in the DSC scans probably melt distinctly apart from each other. The reversal in melting peak intensities on increased elongation leads one to speculate that the thermodynamically less stable, lower melting structures form during mechanical deformation at the expense of the higher melting crystals.

It is well known that the extent to which isothermal crystallization in polymers occurs can be profoundly influenced by molecular orientation. Recently Nakamura et al. (275) have performed a detailed theoretical analysis of the relationships between crystallization temperature, crystallinity, orientation and cooling conditions for nonisothermal processes. These investigators have also evaluated the crystallization rates under molecular orientation during melt spinning experiments and

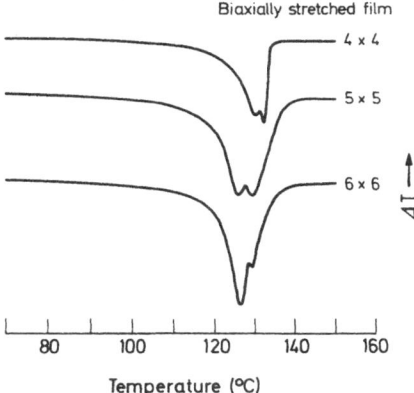

Fig. 47. DSC scans for biaxially stretched linear polyethylene. (Note the increase in peak area for the lower melting peak with increasing stretch ratio.) [Ref. (*273*)]

present an estimation of the increase in rate of crystallization under molecular orientation for very high speed spinning of PET.

5. Effect of Pressure — Undercooling

As already mentioned in Section D.1g, high molecular weight polymers will crystallize into extended chains only by the application of very high pressures. Extended chain crystals are of special importance in that they are nearly completely crystalline and serve as standards for the determination of equilibrium properties of polymers. Recent studies (*262*) on so-called extended chain polyethylene and polyoxymethylene have shown that annealing these crystals even for short periods of time results in "re" folding, a phenomena which is of interest with respect to the thermal stability of these superstructures and also in clarifying the origin of chain folding during melt crystallization. While the causes for "re" folding are not yet known, it has been suggested (*276*) that the lateral surface free energy σ_e in effect lowers the melting point of extended chain crystals, a process which is followed by new crystallization of the melt into chain folded lamellae. The quantity σ_e has very recently been subject of a study by Calvert and Uhlmann (*277*) in an attempt to interpret the high pressure crystalli-

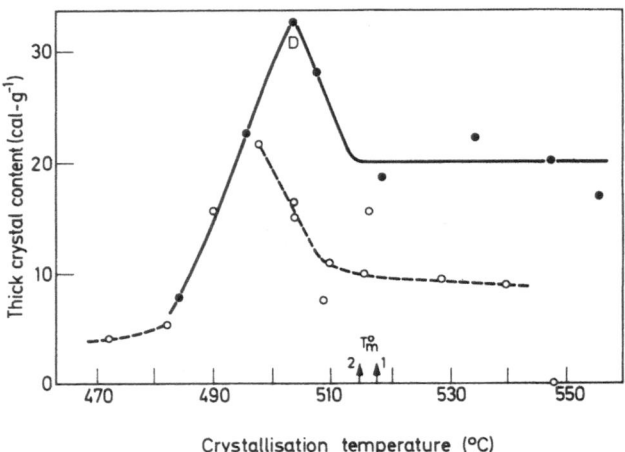

Fig. 48. Variation of thick crystal content with crystallization temperature for Marlex 6009 polyethylene: (O) crystallized 30 min at 5.18 kb, cooled at 17 °K/min; (●) crystallized 35 min at 5.0 kb, cooled at 2 °K/min. D denotes double peak is DSC trace. [Ref. (277)]

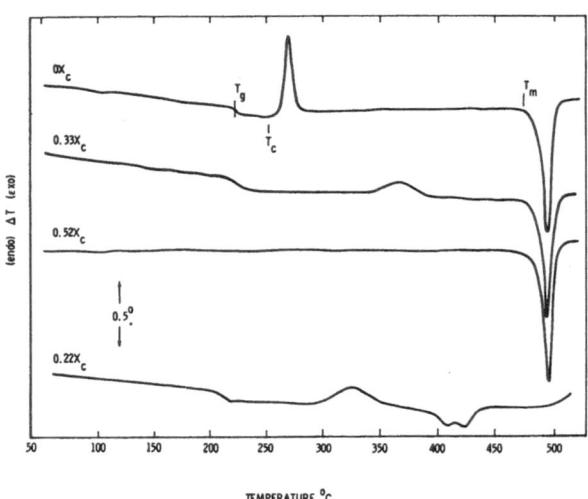

Fig. 49. Differential scanning calorimetry of isotropic amorphous poly (2.6-diphenyl)1.4-phenylene ether (OX) and of varying degree of crystallinity and crystal perfection. [Ref. (102)]

zation kinetics in polyethylene. From the dependence of lamellar thickness on the crystallization undercooling at 5 kb (see Fig. 48), it was determined that the lamellar surface energy (e.g., which controls the lamellar thickness) surprisingly increases with pressure leading to the formation of very thick (but not extended chain) crystals at high pressures.

Although the DSC method used by the authors to separate the thick and thin crystals obtained at high pressures was somewhat arbitrary, it did provide useful means to measure the (Switchboard or folded chain lamellae) type of crystals present in their study. The formation of extended chain crystals was eliminated by utilizing relatively high cooling rates. From the results such as those illustrated in Fig. 48 it was concluded that 1. there is no evidence to support the growth-thickness mechanism as discussed in Section E.1 for thick crystal formation at high pressures; 2. except under conditions of very slow cooling, extended chain crystals do not form with high molecular weight polymer at high pressures; and 3. the increase in lamellar thickness at high pressures cannot be attributed to high annealing rates; the growth process here is similar to that at ordinary pressures.

6. Cold Crystallization

Whenever one observes a crystallization exotherm on heating bulk polymer at temperatures far below the melting point and above the glass transition temperature, the process is called cold crystallization. Cold crystallization has been observed at temperatures as low as $-100°$ C for polysiloxanes (38), and as high as $350°$ C for partially crystalline poly (2.6-diphenyl)phenylene oxide (102) (See Fig. 49 curve 0.33 X_c).

At the present time, nothing is known about the mechanisms of cold crystallization, even though the process is observed in such important commercial materials as polyethylene terephthalate and polyurethanes (38). It is generally assumed that cold crystallization occurs without massive molecular rearrangements and leads to the production of imperfectly formed crystalline regions (38). In contrast, the author has, as previously mentioned, evidence that for crystallizable aromatic polymers, cold crystallization leads to the formation of large and relatively thick "single" polymer crystals (102)

F. References

1. Nernst, W.: Ann. Physik **36**, 395 (1911).
2. Southard, J. C., Brickwedde, F. G.: J. Am. Chem. Soc. **63**, 3142 (1941).
3. Scott, R. B., Meyers, C. H., Rands, R. D., Brickwedde, F. G., Beckedahl, N.: J. Res. Natl. Bur. Std. **35**, 39 (1945).
4. Dole, M., Hettinger, W. P., Larson, N., Wethington, J. A., Worthington, A. E.: Rev. Sci. Inst. **22**, 11, 812 (1951).
5. Worthington, A. E., Marx, P. C., Dole, M.: Rev. Sci. Instr. **26**, 7, 698 (1955).
6. Wilhoit, R. C., Dole, M.: J. Phys. Chem. **57**, 14 (1953).
7. West, E. D., Ginnings, D. C.: J. Res. Natl. Bur. Std. **60** (4), 2848 (1958).
8. Warfield, R. W., Petree, M. C., Donovan, P.: SPE Journal **15**, 12 (1959).
9. Hellwege, K. H., Knappe, W., Semjanow, V.: V. Angew. Phy. **11**, 285 (1959).
10. Sturtevant, J. M.: In: Weissberger (Ed.): Techniques of Org. Chem. 3rd Ed., Vol. 1, Ch. X. New York: Interscience Publ. 1959.
11. Tunnicliff, D. D., Badley, J. H.: Rev. Sci. Instr. **31**, 953 (1960).
12. Bartenev, G. M., Gobatkina, Yu. A., Lukyanov, I. A.: Plasticheskie Massy. 56 (1963).
13. Passalgia, E., Kevorkian, H. K.: J. Appl. Polymer Sci. **7**, 119 (1963).
14. Karaz, F. E., O'Reilly, J. M.: Rev. Sci. Instr. **37**, 255 (1966).
15. Tucker, J. E., Reese, W.: J. Chem. Phys. **46**, 1388 (1967).
16. Pavlinov, I. B., Rabinovich, A. N., Oklandnov, Arzhakov, S. A.: Polymer Sci. USSR **9**, 539 (1967).
17. Melia, T. P., Tyson, A.: Makromol. Chem. **109**, 87 (1967).
18. Grewer, Thv., Wilski, H.: Kollid-Z. u. Z. Polymere **229**, 137 (1969).
19. Burdazhanadze, T. V., Privalov, P. L., Tarvkhelidze, N. N.: Bull. Acad. Sci. Georgian SSR **31**, 277 (1963).
20. Göritz, D., Müller, F. H.: Kolloid-Z. u. Z. Polymere **241**, 1075 (1970).
21. Hager, N. E.: Rev. Sci. Instr. **35**, 618 (1964).
22. Dole, M.: Fortschr. Hochpolymer.-Forsch. **2**, 222 (1960).
23. Brandrup, J., Immergut, E. H.: Polymer Handbook (Interscience Publ.), VI − 61 (1965).
24. Meada, Y., Kanetsuna Hisaaki: Sen-i Gakkaishi **26**, 61 (1970).
25. Feher, F., Görler, G. P.: Z. Angew. Phys. **31**, 53 (1971).
26. Kiziriya, E. L., Roinishoili, E. Yu.: Prib. Tekh. Eksp. **1**, 241 (1971).
27. Bachman, R., DiSalvo, F. J., Geballe, T. H., Greene, R. C., Howard, R. E., King, C. N., Kirsch, H. C., Lee, K. N., Schwall, R. E., Thomas, H. U., Zubeck, R. B.: Rev. Sci. Inst. **43**, 205 (1972).
28. Niesen, H.: Kunststoff-Rundsch. **6**, 1 (1959).
29. Becker, F., Spalink, F.: Z. Physik. Chem. **26**, 1 (1960).
30. Barskii, Yu. P.: Trudy Cosudarst. Vsesajuz. Nauch. Issledovatel. Inst. Keram. **15**, 159 (1960).
31. Kritskaya, D. A., Larin, I. K., Ponomatev, A. N., Talroze, V. L.: Vysokomolekul. Soedin 6, 1044 (1964).
32. Barkalov, I. M., Goldanskii, V. I., Rapoport, V. B.: Dokl. Akad. Nauk SSSR, 161, 1368 (1965).

94 References

33. Chil-Gevorgyan, G. M., Bonetskaya, A. K., Skuratov, S. M.: Zh. Fiz. Khim. **39**, 1794 (1965).
34. Riedel, O., Wittmer, P.: Makromol. Chem. **97**, 1 (1966).
35. Frisch, M. A., Mackle, H.: J. Sci. Instr. **42**, 186 (1965).
36. Anderson, H. M.: J. Polymer Sci. A-1, **7**, 2889 (1969).
37. Dworkin, A.: European Polymer. J. **7**, 671 (1971).
38. Ke, B.: In: Newer methods of polymer characterization, Chapter IX, Polymer Rev. Vol. 6, pp. 347—445.
39. Wunderlich, B.: In: Weissberger, Rossiter, B. W. (Eds.): Differential thermal analysis, Vol. 1, Part V, Chapter 8. Phys. Methods of Chemistry 1971.
40. Smothers, W. J., Chiang, Y.: Handbook of differential thermal analysis, New York: Chem. Publ. Comp., Inc. 1966.
41. Mackenzie, R. C.: The differential thermal analysis of clays. London: Mineralogical Society, 1957.
42. Strella, S. J.: Appl. Polymer Sci. **7**, 569 (1963).
43. Duval, C.: In: Schwenker, R. F., Garn, P. D. (Eds.): Thermal analysis, Vol. 1, p. 3. Academic Press. N. Y., 1969.
44. Fisher, E. W., Hindrichsen, G.: Kolloid-Z. u. Z. Polymere **213** (1), 93 (1966).
45. Wunderlich, B., Baur, H.: Advanc. Polymer Sci. **7** (2), 260 (1970).
46. Müller, F. H., Martin, H.: Kolloid-Z. **172** (2), 97 (1960).
47. Hoffman, J. D.: J. Am. Chem. Soc. **74**, 1696 (1952).
48. Watson, E. S., O'Neill, M. J., Justin, J., Brenner, N.: Anal. Chem. **36** (7), 1233 (1964).
49. O'Neill, M. J.: Anal. Chem. **36** (7), 1238 (1964).
50. Brenner, N., Bartolot, V. J.: Plastics Technology **1964**, 56.
51. Barrall, E. M., Gernert, J. F., Porter, R. S., Johnson, J. F.: Anal. Chem. **35** (12), 1837 (1963).
52. O'Neill, M. J.: Anal. Chem. **38** (10), 1331 (1966).
53. Berg, L. G., Egunov, V. P.: J. Thermal Anal. **1**, 5 (1969).
54. Strella, S., Erhard, P. F.: J. Appl. Polymer Sci. **13**, 1373 (1969).
55. David, D. J.: J. Thermal Anal. **3**, 247 (1971).
56. Miller, G. W.: J. Appl. Polymer Sci. **15**, 2335 (1971).
57. Gobrecht, H., Hamann, K., Willers, G.: J. Phys. E **4**, 21 (1971).
58. Pella, E., Neburoni, M.: J. Thermal Anal. **3**, 229 (1971).
59. Fisher, E. W., Hinrichsen, G.: Kolloid-Z. u. Z. Polymere **213**, 11, 93 (1966); **247**, 858 (1971).
60. Duswalt, A. A.: Plastic Eng. Techn. Pap. **17**, 223 (1971).
61. Gray, A. P.: Inst. News **20**, 10 (1969).
62. Luebke, H. W., Emery, E. M.: Inst. News **20** (1), 6 (1970).
63. Otto, G. W., Jordan, K. C.: Sci-Abstr. **25** (12), 26548 (1972).
64. Ozawa, T., Jsozaki, H., Negishi, A.: Thermochim. Acta **1970**, 545.
65. Wendland, W. W., Bradley, W. S.: Anal. Chim Acta **397**, 52 (1970).
66. Wunderlich, B.: J. Phys. Chem. **69**, 2078 (1965).
67. Wrasidlo, W.: J. Polymer Sci. A-2, **9**, 1603 (1971).
68. Bekkedahl, N.: J. Res. Natl. Bur. Std. **42**, 145 (1949).
69. Rubens, L. C., Skochdopole, R. E.: Encycl. Polym. Sci. Technol. **5**, 83 (1966).
70. Murphy, C. B.: Anal. Chem. **40**, 380 R (1968).
71. Schwenker, R. F.: In: Proc. 2nd Toronto Symp. Thermal Anal. p. 59, 1967.
72. Neimark, B. E., Brodskii, B. R.: Zavodsk Lab. **32**, 1154 (1966).
73. Evans, D. J., Winstanley, C. J.: J. Sci. Instr. **43**, 772 (1966).
74. Harrison, E. A., Wilkes, P.: J. Phys. E. **5**, 174 (1972).
75. Tung, L. H.: J. Polymer Sci. A-2, **5**, 391 (1967).

76. Tsikis, D. S., Polyakov, E. V.: Zh. Fiz. Khim. **40**, 2651 (1966).
77. Hara, K., Schonhorn, H.: J. Appl. Polymer Sci. **16**, 1103 (1972).
78. Miller, G. W.: J. Appl. Polymer Sci. **15**, 2335 (1971).
79. — J. Appl. Polymer Sci. **15**, 1985 (1971).
80. Niezette, J., Desreux, V.: J. Appl. Polymer Sci. **15**, 1981 (1971).
81. McKinney, J. E., Penn, R. W.: Rev. Sci. Instr. **43**, 1211 (1972).
82. Vocks, J. F., Crane, R. A.: Anal. Chem. **31**, 1906 (1959).
83. Duvdevani, I. J., Biesenberger, J. A., Gogas, C. G.: Soc. Plas. Eng. Ann. Tech. Conf. **28**, 110 (1970).
84. Miller, R. P., Sommer, G.: J. Sci. Instr. **43**, 293 (1966).
85. Wilburn, F. W., Metcalfe, S. A., Warburton, R. S.: Glass Technol. **6**, 107 (1965).
86. Dixon, G. D.: J. Phys. E, 4 (1971).
87. Mettler Instrument Corp., Princeton, N. Y., Tech. Bull. FP-1.
88. Kohl, R. W.: U. S. Atomic Energy Commission Dept. MLM-1271, Mosanto Res. Corp., June 1965.
89. Barral, E. M., Gallegos, E. J.: J. Polymer Sci. A-2, **5**, 113 (1967).
90. Perinet, G.: Bull. Soc. France, Mineral. Crist. **89**, 325 (1966).
91. Tsvetkov, V. N.: In: Polymer Rev., Vol. 6, Chapter XIV, p. 563 (1964).
92. Stein, R. S.: Ann. N. Y. Akad. Sci. **155**, 566 (1969).
93. Jackson, J. B., Longman, G. W.: Polymer **10**, 11 (1969).
94. Clough, S., Rhodes, M. P., Stein, R. S.: J. Polymer Sci. C **18**, (1967).
95. Grosius, P., Gallot, Y., Skoulios, A.: Macromol. Chem. **136**, 191 (1970).
96. Saotome, K., Komoto, H.: J. Polymer Sci. A-1, **5**, 107 (1967).
97. Kovacs, A. J., Hobbs, S. Y.: J. Appl. Polymer Sci. **16**, 301 (1972).
98. Shulz, A. R., Gendron, B. M.: J. Appl. Polymer Sci. **16**, 461 (1972).
99. Arakawa, T., Wunderlich, B.: J. Polymer Sci. A-2, **4**, 53 (1966).
100. Wrasidlo, W.: J. Polymer Sci. A-2, **10**, 1719 (1972).
101. Duwez, P.: Trans. AIME **191**, 765 (1951).
102. Wrasidlo, W.: Macromolecules **4**, 642 (1971).
103. Pietrokowsky, P.: J. Sci. Instr. **34**, 445 (1962).
104. Wunderlich, B.: Kolloid-Z. u. Z. Polymer. **231**, 605 (1969).
105. Petrie, S. E. B.: J. Polymer Sci. A-2, **10**, 1255 (1972).
106. Wunderlich, B., Cormier, C. M., Keller, A., Machin, M. J.: J. Macromol. Sci. **3** (1), 93 (1967).
107. — Bodily, D. M.: J. Polymer Sci. C **6**, 137 (1964).
108. — Thermochimica Acta **4**, 175 (1972).
109. Sharonov, Yu. A., Volkenstein, M. V.: In: Porai-Koshits, E. A. (Ed.): The structure of glasses, Vol. 6, Consultants Bureau, N. Y., 1966, and Volkenstein, M. V., Sharonov, Y. Y.: Vyskomolekul **4**, 917 (1962).
110. Frank, W., Stuart, H. A.: Kolloid-Z. u. Z. Polymer. **225**, 1 (1968).
111. Eldridge, J. E.: J. Appl. Polymer Sci. **11**, 1199 (1967).
112. Kastner, S.: J. Polymer Sci. C **16**, 4121 (1968).
113. Bartenev, G. M., Yu, V.: J. Polymer Sci. C **16**, 3591 (1968).
114. Kovacs, A. J.: Fortschr. Hochpolymer.-Forsch. **3**, 394 (1966).
115. Okamoto, H.: J. Polymer Sci. A-2, **8**, 311 (1970).
116. Geil, P. H.: Single crystals. New York: Interscience 1963.
117. Rijke, A. M., Mandelkern, L.: J. Polymer Sci. A-2, **8**, 225 (1970).
118. Pelzbauer, Z., John Manley, R. St.: J. Polymer Sci. A-2, 649 (1969).
119. Erä, V. A., Lindberg, J. J.: J. Polymer Sci. A-2, **10**, 937 (1972).
120. Shen, M. E., Eisenberg, A.: Progress in solid state chemistry, Vol. 3, 407 (Oxford, 1966).
121. Kanig, G.: Kolloid-Z. u. Z. Polymer. **233**, 54 (1969).

122. Boyer, R. F.: Rubber Chem. Technol. **36**, 1303 (1963).
123. Ferry, J. D.: Viscoelastic properties of polymers. New York: J. Wiley and Sons, Inc. 1969.
124. Enskog, D.: Kgl. Svenska Vetenskapsakad. Handl. **63** (4), 1922.
125. Hirai, N., Eyring, H.: J. Appl. Phys. **29**, 810 (1958).
126. Frenkel, J. I.: Kinetische Theory der Flüssigkeiten, 200 (Berlin 1957).
127. Fox, T. G., Flory, P. J.: J. Appl. Phys. **21**, 581 (1950); — J. Polymer Sci. **14**, 315 (1954).
128. Hirai, N., Eyring, H.: J. Polymer Sci. **37**, 51 (1959).
129. Kovacs, A. J.: Thesis presentées à la faculté des science de l'université de Paris, 1955.
130. Someynsky, T., Simha, R.: J. Appl. Phys. **42**, 4545 (1971).
131. Williams, M. L., Landel, R. F., Ferry, J. D.: J. Am. Chem. Soc. **77**, 3701 (1955).
132. Simha, R., Boyer, R. F.: J. Chem. Phys. **37**, 1003 (1962).
133. — Weil, C. E.: J. Macromol. Sci. Phys. B **4**, 215 (1970). — Boyer and Simha: J. Polymer Sci. B **11**, 33 (1973).
134. Sharma, S. C., Mandelkern, L., Stehling, F. C.: J. Polymer Sci. B **10**, 345 (1972).
135. Miller, A. A.: J. Polymer Sci. A-2, 1095 (1964).
136. Cohen, M. H., Turnbull, D.: J. Chem. Phys. **31**, 1164 (1959); **34**, 120 (1961).
137. Naghizadeh, J.: J. Appl. Phys. **35**, 1162 (1964).
138. Chung, H. S.: J. Chem. Phys. **44**, 1362 (1966).
139. Macedo, P. B., Litovitz, T. A.: J. Chem. Phys. **42**, 245 (1965).
140. Turnbull, D., Cohen, M. H.: J. Chem. Phys. **00**, 52 (1970).
141. Gibbs, J. H., DiMarzio, E. A.: J. Chem. Phys. **28**, 373 (1958).
142. Ehrenfest, P.: Leiden Comm. Suppl. 756 (1933), also, see Smolurchowsky, R.: In: Handbook of physics, 2nd Ed., 8—99, McGraw-Hill (1968).
143. Adam, G., Gibbs, J. H.: J. Chem. Phys. **43** (1), 139 (1965).
144. Bestul, A. B., Chang, S. S.: J. Chem. Phys. **40**, 731 (1964).
145. Passaglia, E., Kevorkian, H. K.: J. Appl. Polymer Sci. **7**, 119 (1963). — J. Appl. Phys. **34**, 90 (1963).
146. Tisza, L.: Phase transformations in solids. New York: Wiley 1951.
147. Überreiter, K.: Kolloid-Z. u. Z. Polymere **216**, 217 (1967).
148. McKinney, P. V., Foltz, R. C.: J. Appl. Polymer Sci. **11**, 1189 (1967).
149. Kashmiri, M. I., Sheldon, R. P.: J. Polymer Sci. B **7**, 51 (1969).
150. Bruns, W., Mehdorn, F., Überreiter, K.: Kolloid-Z. u. Z. Polymere **244**, 204 (1971).
151. Breuer, H., Rehage, G.: Kolloid-Z. u. Z. Polymere **216**, 159 (1967).
152. Wunderlich, B., Bodily, D. M., Kaplan, M. H.: J. Appl. Phys. **35** (1), 95 (1964).
153. Miller, A. A.: J. Polymer Sci. A-2, **4**, 415 (1966).
154. Anderson, O. L.: Progress in solid state chemistry, **3**, 179 (Oxford 1966).
155. Karaz, F. E., Bair, H. E., O'Reilly, J. M.: Polymer **8**, 547 (1967).
156. — — J. Phys. Chem. **69**, 2607 (1965).
157. Wunderlich, B.: J. Chem. Phys. **37**, 2429 (1962).
158. Wolpert, S. M., Weitz, A., Wunderlich, B.: J. Polymer Sci. A-2, **9**, 1887 (1971).
159. O'Reilly, J. M., Karaz, F. E.: J. Polymer Sci. C **14**, 49 (1966).
160. Wrasidlo, W.: J. Polymer Sci. (to be published).
161. Wunderlich, B.: J. Phys. Chem. **64**, 1052 (1960).
162. Bunn, C. W.: J. Polymer Sci. **16**, 323 (1955).
163. Petuhkov, B. V.: Techn. Polyester Fibers, p. 32. New York: MacMillan Co. 1963.

164. Kolb, H. J., Izard, E. F.: J. Appl. Phys. **20**, 564 (1949).
165. Brandrup, J., Immergut, E. H.: Polymer Handbook, ———: Interscience 1965.
166. Boyer, R. F., Spencer, R. S.: J. Appl. Phys. **15**, 398 (1944).
167. Klarman, A. F., Galanti, A. V., Sperling, L. H.: J. Polymer Sci. **7**, 1513 (1969).
168. Kryszewsky, M., Mucha, M.: J. Appl. Polymer Sci. **15**, 2687 (1971).
169. Matsuoka, S., Ischida, Y.: J. Polymer Sci. C **14**, 257 (1966).
170. Hoffmann, J. D., Williams, G., Passaglia, E.: J. Polymer Sci. C **14**, 209 (1966).
171. Heijboer, J.: Proc. Intern. Conf. on Phys. of Noncrystalline Solids, p. 231. Ed. by J. A. Prins. North Holland Publ., 1965.
172. Crist, B., Peterlin, A.: J. Polymer Sci. A-2, **9**, 557 (1971).
173. Ishida, Y., Matsuo, M., Ueno, Y., Takayanagi, M.: Kolloid-Z. **199**, 69 (1964).
174. Wrasidlo, W.: J. Macromol. Sci. B (6), **3**, 559 (1972).
175. Batchinski, A. J.: Z. Phys. Chem. **84**, 644 (1913).
176. Doolittle, A. K., Doolittle, D. B.: J. Appl. Phys. **28**, 901 (1957).
177. Bueche, F.: In: Physical properties of polymers. ———: Interscience Publ. (1962).
178. Mercier, J. P., Aklonis, T. J., Litt, M., Tobolsky, A. V.: J. Appl. Polymer Sci. **9**, 447 (1965).
179. Zachmann, H. G.: Fortschr. Hochpolymer. **3**, 481 (1964).
180. Wunderlich, B.: Ber. Bunsenges. Phys. Chem. **74**, 768 (1970).
181. Miller, R. L.: In: Polymer Handbook, Vol. III, p. 32. ———: Interscience Publ. (1967).
182. Jaffe, M., Wunderlich, B.: Kolloid-Z. u. Z. Polymere **216**, 203 (1966).
183. Miyagi, A., Wunderlich, B.: J. Polymer Sci. A-Z 1401 (1972).
184. Wunderlich, B., Sullivan, P., Arakawa, T., Dicyan, A. B., Flood, J. F.: J. Polymer Sci. A **1**, 3581 (1963).
185. Alfrey, T., Mark, H.: J. Phys. Chem. **46**, 112 (1942).
186. Zachmann, H. G.: Kolloid-Z. u. Z. Polymere **216—217**. 180 (1967); **213**, 39 (1966); **231**, 504 (1969).
187. Fisher, E. W.: Kolloid Z. u. Z. Polymere **218**, 97 (1967).
188. Roe, R. J., Smith, K. J., Krigbaum, W.: J. Chem. Phys. **35**, 1306 (1961).
189. Schrader, E., Zachmann, H. G.: Kolloid-Z. u. Z. Polymere **241**, 1015 (1970).
190. Illers, K. H., Hendus, H.: Kolloid-Z. u. Z. Polymere **218**, 56 (1967).
191. Mandelkern, L., Fatou, J. G., Denison, R., Justin, J.: J. Polymer Sci. B **3**, 803 (1965).
192. Flory, P. J., Vrij, A.: J. Am. Chem. Soc. **85**, 3548 (1963).
193. Billmeyer, F. W.: J. Appl. Phys. **28**, 1114 (1957).
194. Broadhurst, M. G.: J. Natl. Bur. Std. **66**, 241 (1962).
195. Wunderlich, B., Cormier, C. M.: J. Polymer Sci. A-2, **5**, 987 (1967).
196. Keith, H. D., Padden, F. J.: J. Appl. Phys. **35**, 1270 (1964).
197. Anderson, F. R.: J. Appl. Phys. **35**, 64 (1964).
198. Pennings, A. J., Kiel, A. M.: Kolloid-Z. **205**, 160 (1965).
199. Huseley, T. W., Bair. H. E.: J. Polymer Sci. B **5**, 265 (1967).
200. Monobe, K., Yamashita, Y., Fujiwara, Y.: Mem. School Eng. Okayama Univ. **3**, 77 (1968).
201. Rijke, A. M., Mandelkern, L.: J. Polymer Sci. A-2, **8**, 225 (1970).
202. Geil, P. H., Anderson, F. R., Wunderlich, B., Arakawa, T.: J. Polymer Sci. A-2, 3707 (1964).
203. Rees, D. V., Bassett, D. C.: J. Polymer Sci. B **7**, 273 (1969).
204. Parks, W., Richards, R. B.: Trans. Faraday Soc. **45**, 203 (1949).
205. Matsuoka, S.: J. Polymer Sci. **42**, 511 (1960).
206. Wunderlich, B.: J. Polymer Sci. A **2**, 3697 (1964).

207. Southern, J. H., Porter, R. S., Bair, H. E.: J. Polymer Sci. A-2, **10**, 1135 (1972).
208. Davidson, T., Wunderlich, B.: J. Polymer Sci. A-2, **7**, 377 (1969).
209. Kim, H., Mandelkern, L.: J. Polymer Sci. A-2, **10**, 1125 (1972).
210. Lemstra, P. J., Kooistra, T., Challa, G.: J. Polymer Sci. A-2, **10**, 823 (1972).
211. Krigbaum, W. R., O'Mara, J. H.: J. Polymer Sci. A-2, **8**, 1011 (1970).
212. Lovering, E. G., Wooden, D. C.: J. Polymer Sci. A-2, **9**, 175 (1971).
213. Braun, W., Hellwege, K. H., Knappe, W.: Kolloid-Z. **215**, 10 (1967).
214. Mandelkern, L., Yain, N. L., Kim, H.: J. Polymer Sci. A-2, **6**, 165 (1968).
215. Hay, J. N., Sabir, M., Stephen, R. L. T.: Polymer **10**, 187 (1969).
216. Beech, D. R., Booth, C.: J. Polymer Sci. B, **8**, 731 (1970).
217. Allen, G., Booth, C., Jones, M. N., Marks, D. J., Taylor, W. D.: Polymer **5**, 547 (1964).
218. Cooper, W., Eaves, D. G., Vaughan, G.: Polymer **8**, 273 (1967).
219. Aggarwal, S. L., Marker, L., Kollar, W. L., Geroch, R.: Advan. Chem. Ser. **52**, 88 (1966).
220. Booth, C., Devey, C. J., Dodgson, D. V., Hillier, I. H.: J. Polymer Sci. A-2, **8**, 519 (1970).
221. Ke, B., Sisko, A. W.: J. Polymer Sci. **50**, 87 (1961).
222. Liberti, F. N., Wunderlich, B.: J. Polymer Sci. A-2, **6**, 833 (1968).
223. Busch, M., Weber, W.: J. Prakt. Chem. **146**, 1 (1936).
224. Rothe, M.: In: Brandrup and Immergut (Ed.): Polymer Handbook. Interscience Publ., VII–1 (1965).
225. Wrasidlo, W.: J. Polymer Sci. A-2, **10**, 437 (1972).
226. Shulz, A. R., McCullough, R. C.: J. Polymer Sci. A-2, **7**, 1577 (1969); **10**, 307 (1972).
227. Shulz, A. R.: J. Polymer Sci. A-2, **8**, 883 (1970).
228. Packter, A., Sharif, K. A.: J. Polymer Sci. B **9**, 435 (1971).
229. Karaz, F. E., Mangaraj, D.: ASC Polymer Reprints **12** (1), 317 (1971).
230. Mandelkern, L.: Chem. Revs. **56**, 903 (1956).
231. Dole, M., Wunderlich, B.: Makromol. Chem. **34**, 29 (1959).
232. Izard, E. F.: J. Polymer Sci. **8**, 503 (1952).
233. Carothers, W. H.: Chem. Revs. **8**, 353 (1931).
234. Hill, R., Walker, E. E.: J. Polymer Sci. **3**, 609 (1948).
235. Taylor, G. W.: Polymer **3**, 543 (1962).
236. Ishibashi, M.: Polymer **5**, 305 (1964).
237. Hobbs, S. Y., Billmeyer, F. W. Jr.: J. Polymer Sci. A-2, **8**, 1387 (1970); **8**, 1395 (1970).
238. Sweet, G. E., Bell, J. P.: J. Polymer Sci. A-2, **10**, 1273 (1972).
239. Roberts, R. C.: J. Polymer Sci. B, **8**, 381 (1970).
240. Ikeda, M.: Kobimshi Kageku, **25**, 87 (1968).
241. Holdsworth, P. A., Turner-Jones, A.: Polymer **12**, 195 (1971).
242. Dewar, M. J. S., Goldberg, R. S.: J. Am. Chem. Soc. **92**;6, 1582 (1970).
243. Flory, P. J.: J. Chem. Phys. **17**, 223 (1949).
244. Hoffmann, J. D., Lauritzen, J. I. Jr., Passaglia, E., Ross, G. S., Frolen, L. J., Weeks, T. J.: Kolloid-Z. **231**, 564 (1969).
245. Zachmann, H. G.: Kolloid-Z. u. Z. Polymere **231**, 504 (1969).
246. Dorman, R. H., Roberts, B. W., Turnbull, D. (Ed.): Mandelkern, L.: Growth and perfection of crystals, John Wiley Publ., (1958); also in Crystallization of Polymers. New York: McGraw-Hill 1964.
247. Kilian, H. G.: Kolloid-Z. u. Z. Polymere **231**, 534 (1969).
248. Calvert, P. D., Uhlmann, D. R.: J. Appl. Phys. **43**, 944 (1972).
249. Binsbergen, F. L.: Kolloid-Z. u. Z. Polymere **238**, 389 (1969).

250. Roe, R. J., Smith, K. J., Kingbaum, W.: J. Chem. Phys. **35**, 1306 (1961).
251. Vagfarov, M. Sh.: V. Vysokomol. Soedin. A-**10**, 1267 (1967); A-**11**, 1195 (1969).
252. Hendus, H., Illers, K. H.: Kunststoffe **57** (3), 193 (1967).
253. Hobbs, S. Y., Mankin, G. I.: J. Polymer Sci. A-2, **9**, 1907 (1971).
254. Mandelkern, L., Allou, A. L., Gopalan, M.: J. Phys. Chem. **72**, 309 (1968).
255. Hamada, F., Wunderlich, B., Sumida, T., Hayashi, S., Nakajima, A.: J. Phys. Chem. **72**, 178 (1968).
256. Powers, J., Miller, R. L.: In: Polymer Handbook, p. III-93. ————: Interscience Publ. 1965.
257. Banks, W., Gordon, M., Roe, R. J., Sharples, A.: Polymer **4**, 61 (1963).
258. Keith, H. D.: Kolloid-Z. u. Z. Polymere **231**, 421 (1969).
259. Rybnikar, F. J.: J. Polymer Sci. **44**, 517 (1960).
260. Peterlin, A., Roeckl, E.: J. Appl. Phys. **34**, 102 (1963).
261. Baur, H.: Ber. Bunsenges. Phys. Chem. **71**, 703 (1967).
262. Fisher, E. W.: Kolloid-Z. u. Z. Polymere **231**, 472 (1969).
263. Mandelkern, L., Allou, A. L.: J. Polymer Sci. B-**4**, 447 (1966).
264. Kawai, T.: Kolloid-Z. u. Z. Polymere **201**, 104 (1965).
265. Mandelkern, L., Posner, A. S., Diano, A. F., Roberts, D. E.: J. Appl. Phys. **32**, 1509 (1961).
266. Hoffmann, J. D., Weeks, T. J.: J. Chem. Phys. **42**, 4301 (1965).
267. Hirai, N., Yamashita, Y.: Chem. High Polymer (Japan) **21**, 173 (1964).
268. Gumargalieva, K. F., Belavceva, E. M.: Vysokomol. Soedin. **8**, 1604 (1966).
269. Balta Calleja, F. J., Keller, A.: J. Polymer Sci. A-2, 2171 (1964).
270. Spegt, P., Skoulios, A.: Compt. Rend. Acad. Sci. **262**, 722 (1966).
271. Bonart, R.: Kolloid-Z. u. Z. Polymer **231**, 438 (1969).
272. Ward, I. M. (Ed.): J. Mat. Sci. **6** (6), 451—596 (1971).
273. Adams, G. C.: J. Polymer Sci. A-2, **9**, 1235 (1971).
274. Uejo, Hoshino, S.: J. Appl. Polymer Sci. **14**, 317 (1970).
275. Nakamura, K., Watanabe, T., Katayama, K., Amano, T.: J. Appl. Polymer Sci. **16**, 1077 (1972).
276. Wunderlich, B., Melillo, L.: Science **154**, 1329 (1966).
277. Calvert, P. D., Uhlmann, D. R.: J. Polymer Sci. A-2, **10**, 1811 (1972).

Received July 9, 1973

"Advances in Polymer Science" comprises reports of monograph type dealing with progress achieved in the physics and chemistry of high polymers; it includes full bibliographical data. The objective is to provide those working in this field with information on subjects that are of special interest, and to report recent advances that have been so rapid as to require review-type treatment.

Individual parts of volumes 5-7 can be purchased separately. Volumes 8-11 are not published in separate parts.

■ Prospectus
  on request

**Springer-Verlag
Berlin
Heidelberg
New York**

München Johannesburg London New Delhi Paris Rio de Janeiro Sydney Tokyo Wien

Advances in Polymer Science

Fortschritte der Hochpolymeren-Forschung

Edited by: H.-J. Cantow, G. Dall'Asta, J. D. Ferry, H. Fujita, W. Kern, G. Natta, S. Okamura, C. G. Overberger, W. Prins, G. V. Schulz, W. P. Slichter, A. J. Stavermann, J. K. Stille, H. A. Stuart

Volumes 1-4
Out of print

Volume 5
185 figs. IV, 619 pp. (231 pp. in German). 1967/68
Cloth DM 189,—; US $77.50   ISBN 3-540-04034-X

Volume 6
236 figs. III, 574 pp. (128 pp. in German). 1969
Cloth DM 172,—; US $70.60   ISBN 3-540-04401-9

Volume 7
252 figs. III, 593 pp. (164 pp. in German). 1970/71
Cloth DM 186,—; US $76.30   ISBN 3-540-05342-5

Volume 8
57 figs. III, 237 pp. 1971
Cloth DM 88,—; US $36.10   ISBN 3-540-05483-9

Volume 9
142 figs. III, 414 pp. (227 pp. in German). 1972
Cloth DM 134,—; US $55.00   ISBN 3-540-05484-7

Volume 10
50 figs. III, 194 pp. (138 pp. in German). 1972
Cloth DM 78,—; US $32.00   ISBN 3-540-05838-9

Volume 11
56 figs. III, 204 pp. (93 pp. in German). 1973
Cloth DM 88,—; US $36.10   ISBN 3-540-06054-5

Volume 12
62 figs. III, 190 pp. 1973
Cloth DM 78,—; US $32.00   ISBN 3-540-06431-1

Prices are subject to change without notice

B. Vollmert: Polymer Chemistry

Translator:
E. H. Immergut

With approx. 238 figs.
Approx. 650 pages.
1973. Cloth approx.
DM 94,50;
approx. US $38.80
ISBN 3-540-05631-9

Prices are subject
to change
without notice

A comprehensive, fundamental treatment of the formation of synthetic and natural macromolecules, their chemical reactions and their structural characteristics. Special consideration is given to the relationship between synthesis, structure, and properties.

Contents: Structural Principles: Chain Structure, Degree of Polymerization. Copolymers. Branched and Cross-linked Polymers. — Synthesis and Reactions of Macromolecular Compounds: Synthesis of Polymers with Carbon — Carbon Chains through Polymerization of Olefinic Unsaturated Compounds. Synthesis of Macromolecules with Heteroatoms in the Chain. Enzymatic Syntheses. Purification of Polymers. Chemical Transformation of Polymers. — The Properties of the Individual Macromolecule: Molecular Weight. The Form of the Molecules (Molecular Shape). — States of Macromolecular Aggregation: Intermolecular Forces and Aggregation (Association). The Macromolecular Solution. The Gel State. The Rubberelastic State (Viscoelasticity). The Solid State. — Bibliography. — Subject Index.

Springer-Verlag
Berlin   Heidelberg   New York